T0227402

Problems and Solutions in Quantum Physics

Problems and Solutions in Quantum Physics

Zbigniew Ficek

PAN STANFORD PUBLISHING

Published by

Pan Stanford Publishing Pte. Ltd.
Penthouse Level, Suntec Tower 3
8 Temasek Boulevard
Singapore 038988

Email: editorial@panstanford.com
Web: www.panstanford.com

British Library Cataloguing-in-Publication Data
A catalogue record for this book is available from the British Library.

Problems and Solutions in Quantum Physics

ISBN 978-981-4669-36-8 (Hardcover)
ISBN 978-981-4669-37-5 (eBook)

Preface

This book contains problems with solutions of a majority of the tutorial problems given in the textbook *Quantum Physics for Beginners*. Not presented are solutions to only those problems whose solutions the reader can find in the textbook. You should read the text of a chapter before trying the tutorial problems in the chapter. Solutions to the problems give the reader a self-check and reassurance on the progress of learning.

Zbigniew Ficek
The National Centre for Applied Physics
King Abdulaziz City for Science and Technology
Riyadh, Saudi Arabia
Spring 2016

Chapter 1

Radiation (Light) is a Wave

Problem 1.2

Using Eq. (1.13) of the textbook, show that

$$\vec{E}_k = -c\vec{\kappa} \times \vec{B}_k, \tag{1.1}$$

which is the same relation one can obtain from the Maxwell Eq. (1.4).

(*Hint:* Use the vector identity $\vec{A} \times (\vec{B} \times \vec{C}) = \vec{B}(\vec{A} \cdot \vec{C}) - \vec{C}(\vec{A} \cdot \vec{B})$.)

Solution

Equation (1.13) of the textbook shows the relation between the directions of the electric and magnetic fields of the electromagnetic wave

$$\vec{B}_k = \frac{1}{c}\vec{\kappa} \times \vec{E}_k, \tag{1.2}$$

where $\vec{\kappa}$ is the unit vector in the direction of propagation of the wave.

By taking a cross product of both sides from the left with the vector $\vec{\kappa}$, we get

$$\vec{\kappa} \times \vec{B}_k = \frac{1}{c}\vec{\kappa} \times (\vec{\kappa} \times \vec{E}_k). \tag{1.3}$$

Problems and Solutions in Quantum Physics
Zbigniew Ficek
Copyright © 2016 Pan Stanford Publishing Pte. Ltd.
ISBN 978-981-4669-36-8 (Hardcover), 978-981-4669-37-5 (eBook)
www.panstanford.com

Next, using the vector identity $\vec{A} \times (\vec{B} \times \vec{C}) = \vec{B}(\vec{A} \cdot \vec{C}) - \vec{C}(\vec{A} \cdot \vec{B})$, we can write the right-hand side of the above equation as

$$\frac{1}{c}\vec{\kappa} \times (\vec{\kappa} \times \vec{E}_k) = \frac{1}{c}\left[\vec{\kappa}(\vec{\kappa} \cdot \vec{E}_k) - \vec{E}_k(\vec{\kappa} \cdot \vec{\kappa})\right]. \tag{1.4}$$

Since $\vec{\kappa} \cdot \vec{\kappa} = 1$ and the electric and magnetic fields are transverse fields $(\vec{\kappa} \cdot \vec{E}_k = 0)$, we arrive at

$$\vec{E}_k = -c\vec{\kappa} \times \vec{B}_k. \tag{1.5}$$

This result for $\vec{E}_k$ and that for $\vec{B}_k$, Eq. (1.2), show that both $\vec{B}_k$ and $\vec{E}_k$ of an electromagnetic wave are perpendicular to the direction of propagation of the wave.

Problem 1.3

Show, using the divergence Maxwell equations, that the electromagnetic waves in vacuum are transverse waves.

Solution

Consider an electromagnetic wave propagating in the z direction. The wave is represented by the electric and magnetic fields of the form

$$\vec{E} = \vec{E}_0 e^{i(\omega t - kz)},$$
$$\vec{B} = \vec{B}_0 e^{i(\omega t - kz)}. \tag{1.6}$$

The propagation of the wave is characterized by the frequency ω and the wave number k.

When calculating divergences $\nabla \cdot \vec{E}$ and $\nabla \cdot \vec{B}$, we get

$$\nabla \cdot \vec{E} = \frac{\partial E_x}{\partial x} + \frac{\partial E_y}{\partial y} + \frac{\partial E_z}{\partial z} = 0 + 0 + \frac{\partial E_z}{\partial z},$$

$$\nabla \cdot \vec{B} = \frac{\partial B_x}{\partial x} + \frac{\partial B_y}{\partial y} + \frac{\partial B_z}{\partial z} = 0 + 0 + \frac{\partial B_z}{\partial z}. \tag{1.7}$$

Since in vacuum $\nabla \cdot \vec{E} = 0$ and $\nabla \cdot \vec{B} = 0$ always in electromagnetism, we have

$$\frac{\partial E_z}{\partial z} = 0 \quad \text{and} \quad \frac{\partial B_z}{\partial z} = 0. \tag{1.8}$$

However, for the electric and magnetic fields of a plane wave,

$$\frac{\partial E_z}{\partial z} = -ikE_z \quad \text{and} \quad \frac{\partial B_z}{\partial z} = -ikB_z. \tag{1.9}$$

Hence, the right-hand sides must be zero, which means that either $k = 0$ or $E_z = 0$ and $B_z = 0$, that both $\vec{E}$ and $\vec{B}$ are transverse to the direction of propagation. Since $k \neq 0$ for a propagating wave, the wave is transverse in both $\vec{E}$ and $\vec{B}$.

Problem 1.4

Calculate the energy of an electromagnetic wave propagating in one dimension.

Solution

Consider a plane electromagnetic wave propagating in the z direction in a vacuum with the electric field polarized in the x direction:

$$\vec{E} = E_0 \sin(\omega t - kz)\hat{i}, \tag{1.10}$$

where $\hat{i}$ is the unit vector in the x direction.

Having $\vec{E}$, we can calculate the magnetic field of the wave using the Maxwell equation

$$\frac{\partial \vec{B}}{\partial t} = -\nabla \times \vec{E}, \tag{1.11}$$

and get

$$\frac{\partial \vec{B}}{\partial t} = -\nabla \times \vec{E} = kE_0 \cos(\omega t - kz)\hat{j}. \tag{1.12}$$

Integrating this equation, we find

$$\vec{B} = kE_0 \int dt \cos(\omega t - kz)\hat{j} = \frac{kE_0}{\omega} \sin(\omega t - kz)\hat{j}. \tag{1.13}$$

Since $k/\omega = 1/c$, we finally obtain

$$\vec{B} = B_0 \sin(\omega t - kz)\hat{j}, \tag{1.14}$$

where $B_0 = E_0/c$.

The energy of the electromagnetic field is determined by the Poynting vector, defined as

$$\vec{U} = \varepsilon_0 c^2 \vec{E} \times \vec{B} = \varepsilon_0 c^2 E_0 B_0 \sin^2(\omega t - kz)\hat{k}. \qquad (1.15)$$

Since $B_0 = E_0/c$, we have

$$\vec{U} = \varepsilon_0 c E_0^2 \sin^2(\omega t - kz)\hat{k}. \qquad (1.16)$$

Then, the average value $\langle U \rangle$ of the magnitude of the Poynting vector is

$$\langle U \rangle = \varepsilon_0 c E_0^2 \langle \sin^2(\omega t - kz) \rangle = \frac{1}{2} \varepsilon_0 c E_0^2, \qquad (1.17)$$

where we have used the fact that $\langle \sin^2(\omega t - kz) \rangle = 1/2$.

Chapter 3

Blackbody Radiation

Problem 3.1

We have shown in Section 3.1 of the textbook that the number of modes in the unit volume and the unit of frequency is

$$N = N_\nu = \frac{1}{V}\frac{dN(k)}{d\nu} = \frac{8\pi\nu^2}{c^3}. \tag{3.1}$$

In terms of the wavelength λ, we have shown that the number of modes in the unit volume and the unit of wavelength is

$$N = N_\lambda = \frac{8\pi}{\lambda^4}. \tag{3.2}$$

Explain, why it is not possible to obtain N_λ from N_ν simply by using the relation $\nu = c/\lambda$.

Solution

The reason is that ν and λ are not linearly dependent on each other. The frequency ν is inversely proportional to λ. Hence,

$$\frac{d\nu}{d\lambda} = -\frac{c}{\lambda^2}. \tag{3.3}$$

Problems and Solutions in Quantum Physics
Zbigniew Ficek
Copyright © 2016 Pan Stanford Publishing Pte. Ltd.
ISBN 978-981-4669-36-8 (Hardcover), 978-981-4669-37-5 (eBook)
www.panstanford.com

Therefore, when going from the frequency space to the wavelength space, we use the chain rule

$$\frac{1}{V}\frac{dN(k)}{d\lambda} = \frac{1}{V}\frac{dN(k)}{dv}\frac{dv}{d\lambda} = -\frac{8\pi v^2}{c^2 \lambda^2}.$$ (3.4)

Then substituting $v = c/\lambda$, we obtain

$$\frac{1}{V}\frac{dN(k)}{d\lambda} = -\frac{8\pi}{\lambda^4}.$$ (3.5)

Chapter 4

Planck's Quantum Hypothesis: Birth of Quantum Theory

Problem 4.2

Using the Planck formula for P_n, show that

(a) The average number of photons is given by

$$\langle n \rangle = \frac{1}{e^x - 1},\tag{4.1}$$

where $x = \frac{\hbar\omega}{k_B T}$, and k_B is the Boltzmann constant.

(b) Show that for large temperatures ($T \gg 1$), the average energy is proportional to temperature, i.e., $\langle E \rangle = k_B T$.

(c) Calculate $\langle n^2 \rangle$ and show that the ratio

$$\alpha = \frac{\langle n^2 \rangle - \langle n \rangle}{\langle n \rangle^2} = 2.\tag{4.2}$$

Solution (a)

Using the Boltzmann formula for P_n and the definition of average, we find

Problems and Solutions in Quantum Physics
Zbigniew Ficek
Copyright © 2016 Pan Stanford Publishing Pte. Ltd.
ISBN 978-981-4669-36-8 (Hardcover), 978-981-4669-37-5 (eBook)
www.panstanford.com

$$\langle n \rangle = \sum_{n=0}^{\infty} n P_n = \frac{\sum_{n=0}^{\infty} n e^{-nx}}{\sum_{n=0}^{\infty} e^{-nx}}, \tag{4.3}$$

where $x = \frac{\hbar\omega}{k_B T}$, and k_B is the Boltzmann constant.

Since the ratio of successive terms in $\sum_{n=0}^{\infty} e^{-nx}$ is a constant,

$$\frac{e^{-(n+1)x}}{e^{-nx}} = e^{-x}, \tag{4.4}$$

the series $\sum_{n=0}^{\infty} e^{-nx}$ is a geometric series of the sum

$$\sum_{n=0}^{\infty} e^{-nx} = \frac{1}{1 - e^{-x}}. \tag{4.5}$$

Hence

$$\langle n \rangle = \sum_{n=0}^{\infty} n P_n = \left(1 - e^{-x}\right) \sum_{n=0}^{\infty} n e^{-nx}. \tag{4.6}$$

We can calculate the sum $\sum_{n=0}^{\infty} n e^{-nx}$ as follows. Denote

$$z = \sum_{n=0}^{\infty} e^{-nx} = \frac{1}{1 - e^{-x}}. \tag{4.7}$$

If we differentiate this expression with respect to x, we obtain

$$\frac{dz}{dx} = -\sum_{n=0}^{\infty} n e^{-nx} = \frac{-e^{-x}}{(1 - e^{-x})^2}. \tag{4.8}$$

Thus, we readily see that

$$\sum_{n=0}^{\infty} n e^{-nx} = \frac{e^{-x}}{(1 - e^{-x})^2}, \tag{4.9}$$

and then

$$\langle n \rangle = \left(1 - e^{-x}\right) \sum_{n=0}^{\infty} n e^{-nx} = \left(1 - e^{-x}\right) \frac{e^{-x}}{(1 - e^{-x})^2} \tag{4.10}$$

$$= \frac{e^{-x}}{1 - e^{-x}} = \frac{e^{-x} e^x}{(1 - e^{-x}) e^x} = \frac{1}{e^x - 1}. \tag{4.11}$$

Solution (b)

Since $E_n = n\hbar\omega$, and using the solution to the part (a), we have

$$\langle E_n \rangle = \langle n \rangle \hbar\omega = \frac{\hbar\omega}{e^x - 1}. \tag{4.12}$$

For large temperatures $T \gg 1$, the parameter $x \ll 1$, and then we can expand the exponent e^x into a Taylor series

$$e^x \approx 1 + x + \ldots, \tag{4.13}$$

from which we find that

$$e^x - 1 \approx x, \tag{4.14}$$

which gives

$$\langle E_n \rangle = \frac{\hbar\omega}{x} = k_B T. \tag{4.15}$$

Thus, for large temperatures, the average energy of a quantum radiation field agrees with that predicted by the Equipartition theorem.

Solution (c)

From the definition of average and using the Boltzmann distribution function, we find

$$\langle n^2 \rangle = \sum_{n=0}^{\infty} n^2 P_n = \left(1 - e^{-x}\right) \sum_{n=0}^{\infty} n^2 e^{-nx}, \tag{4.16}$$

where, as before in (a), $x = \frac{\hbar\omega}{k_B T}$, and k_B is the Boltzmann constant. Since

$$\frac{dz}{dx} = -\sum_{n=0}^{\infty} n e^{-nx} = \frac{-e^{-x}}{(1 - e^{-x})^2}, \tag{4.17}$$

we take a derivative over x of both sides of the above equation and obtain

$$\frac{d^2 z}{dx^2} = \sum_{n=0}^{\infty} n^2 e^{-nx} = \frac{e^{-x}(1 + e^{-x})}{(1 - e^{-x})^3}. \tag{4.18}$$

Hence

$$\langle n^2 \rangle = \frac{e^{-x}(1+e^{-x})}{(1-e^{-x})^2} = \frac{e^{2x}e^{-x}(1+e^{-x})}{e^{2x}(1-e^{-x})^2} \tag{4.19}$$

$$= \frac{(e^x+1)}{(e^x-1)^2}. \tag{4.20}$$

From the relation

$$\langle n \rangle = \frac{1}{e^x - 1}, \tag{4.21}$$

we find that

$$e^x = 1 + \frac{1}{\langle n \rangle}, \tag{4.22}$$

and after substituting this result into Eq. (4.20), we obtain

$$\langle n^2 \rangle = \frac{2\langle n \rangle + 1}{\langle n \rangle} \langle n \rangle^2 = \langle n \rangle + 2\langle n \rangle^2. \tag{4.23}$$

Hence

$$\alpha = \frac{\langle n^2 \rangle - \langle n \rangle}{\langle n \rangle^2} = \frac{\langle n \rangle + 2\langle n \rangle^2 - \langle n \rangle}{\langle n \rangle^2} = 2. \tag{4.24}$$

The parameter α is known in statistical physics as a measure of correlations (distribution) between photons in a radiation field. The value $\alpha = 2$ means that in a thermal field, the correlations between photons are large. In other words, the photons group together (move in large groups). This effect is often called *photon bunching*.

Problem 4.3

Suppose that photons in a radiation field have a Poisson distribution defined as

$$P_n = \frac{\langle n \rangle^n}{n!} e^{-\langle n \rangle}. \tag{4.25}$$

Calculate the variance of the number of photons defined as $\sigma_n = \langle n^2 \rangle - \langle n \rangle^2$ and show that the ratio $\alpha = 1$.

Solution

With the Poisson distribution of photons

$$P_n = \frac{\langle n \rangle^n}{n!} e^{-\langle n \rangle}, \tag{4.26}$$

the average number of photons is given by

$$\langle n \rangle = \sum_n n P_n = \sum_n \frac{n \langle n \rangle^n}{n!} e^{-\langle n \rangle}$$

$$= \langle n \rangle\, e^{-\langle n \rangle} \sum_n \frac{\langle n \rangle^{n-1}}{(n-1)!} = \langle n \rangle, \tag{4.27}$$

where we have used the fact that

$$\sum_n \frac{\langle n \rangle^{n-1}}{(n-1)!} = e^{\langle n \rangle}, \tag{4.28}$$

i.e., the above sum is a Taylor expansion of $e^{\langle n \rangle}$.

Similarly, we can calculate $\langle n^2 \rangle$ as

$$\langle n^2 \rangle = \sum_n n^2 P_n = \sum_n \frac{n^2 e^{-\langle n \rangle} \langle n \rangle^n}{n!} = \langle n \rangle\, e^{-\langle n \rangle} \sum_n \frac{n \langle n \rangle^{n-1}}{(n-1)!}. \tag{4.29}$$

To proceed further with the sum over n, we change the variable by substituting $n - 1 = k$ and obtain

$$\langle n^2 \rangle = \langle n \rangle\, e^{-\langle n \rangle} \left\{ \sum_k \frac{k \langle n \rangle^k}{k!} + \sum_k \frac{\langle n \rangle^k}{k!} \right\}. \tag{4.30}$$

The two sums over k are easy to evaluate, and finally we obtain

$$\langle n^2 \rangle = \langle n \rangle^2 + \langle n \rangle. \tag{4.31}$$

Thus, the variance of photons in a field with the Poisson distributions is given by

$$\sigma_n = \langle n^2 \rangle - \langle n \rangle^2 = \langle n \rangle, \tag{4.32}$$

and then we readily find from the definition of α that

$$\alpha = 1. \tag{4.33}$$

The value of $\alpha = 1$ means that photons in the field with Poisson distribution are independent of each other. Such a field is called a *coherent field*.

Problem 4.5

Show that at the wavelength λ_{max}, the intensity $I(\lambda)$ calculated from Planck's formula has its maximum

$$I(\lambda_{max}) \approx \frac{170\pi (k_B T)^5}{(hc)^4}, \qquad (4.34)$$

where k_B is the Boltzmann constant.

Solution

Consider Planck's formula in terms of wavelength

$$I(\lambda) = \frac{8\pi hc}{\lambda^5 \left(e^{hc/\lambda k_B T} - 1\right)}. \qquad (4.35)$$

Since Planck's formula has its maximum at

$$\lambda_{max} = \frac{hc}{4.9651 k_B T}, \qquad (4.36)$$

we find the maximum value of $I(\lambda_{max})$ as

$$I(\lambda_{max}) = \frac{8\pi hc}{(hc)^5}(4.9651 k_B T)^5 \frac{1}{e^{4.9651} - 1}$$

$$= \frac{\pi (k_B T)^5}{(hc)^4} \frac{8 \times (4.9651)^5}{e^{4.9651} - 1}$$

$$\approx \frac{\pi (k_B T)^5}{(hc)^4} 170 = \frac{170\pi (k_B T)^5}{(hc)^4}. \qquad (4.37)$$

Problem 4.6

(a) Derive the Wien displacement law by solving the equation $dI(\lambda)/d\lambda = 0$.

(*Hint:* Set $hc/\lambda k_B T = x$ and show that dI/dx leads to the equation $e^{-x} = 1 - \frac{1}{5}x$. Then show that $x = 4.956$ is the solution.)

(b) In part (a), we have obtained λ_{max} by setting $dI(\lambda)/d\lambda = 0$. Calculate ν_{max} from the Planck formula by setting $dI(\nu)/d\nu = 0$.

Is it possible to obtain ν_{max} from λ_{max} simply by using $\lambda_{max} = c/\nu_{max}$? Note, ν_{max} is the frequency at which the intensity of the emitted radiation is maximal.

Solution (a)

In the Planck formula

$$I(\lambda) = \frac{8\pi hc}{\lambda^5 \left(e^{hc/\lambda k_B T} - 1\right)},\qquad (4.38)$$

we substitute for

$$\frac{hc}{\lambda k_B T} = x,\qquad (4.39)$$

and obtain

$$I(x) = A\frac{x^5}{(e^x - 1)},\qquad (4.40)$$

where

$$A = \frac{8\pi hc(k_B T)^5}{(hc)^5}\qquad (4.41)$$

is a constant independent of λ.

We find the maximum of $I(x)$ by solving the equation

$$\frac{dI(x)}{dx} = 0.\qquad (4.42)$$

Thus

$$\frac{dI(x)}{dx} = A\frac{5x^4(e^x - 1) - x^5 e^x}{(e^x - 1)^2} = A\frac{x^4[5(e^x - 1) - xe^x]}{(e^x - 1)^2}.$$
$$(4.43)$$

Hence

$$\frac{dI(x)}{dx} = 0 \quad \text{when} \quad x = 0 \quad \text{or} \quad 5(e^x - 1) - xe^x = 0.$$
$$(4.44)$$

The root $x = 0$ is unphysical as it would correspond to $T \to \infty$, so we will focus on the solution to the exponent-type equation, which can be written as

$$e^{-x} = 1 - \frac{1}{5}x.\qquad (4.45)$$

This equation cannot be solved exactly. Therefore, we will apply an approximate method.

One can see that there are two roots of the above equation: $x = 0$ (exact root) and an approximate root $x \approx 5$ (as $e^{-5} \approx 0$). The root $x = 0$ is unphysical as it would correspond to $T \to \infty$, so we will focus on the root $x \approx 5$.

How to estimate the exact root if we know an approximate root?

Let x_0 be close to the exact root of $F(x)$ and let $x_0 + \Delta x$ be the exact root. Then, using a Taylor expansion, we can write

$$F(x_0 + \Delta x) = F(x_0) + F'\Delta x = 0, \qquad (4.46)$$

where

$$F'(x_0) = \frac{dF(x)}{dx}\big|_{x=x_0}. \qquad (4.47)$$

Hence, assuming that x_0 is a root of the equation, the error in the estimation of the exact root is equal to

$$\Delta x = -\frac{F(x_0)}{F'(x_0)}. \qquad (4.48)$$

Let

$$F(x) = 5(e^x - 1) - xe^x. \qquad (4.49)$$

Then

$$F'(x) = 5e^x - e^x - xe^x = (4-x)e^x. \qquad (4.50)$$

Thus, for $x \equiv x_0 = 5$:

$$F(5) = 5(e^5 - 1) - 5e^5 = -5,$$
$$F'(5) = -e^5 = -148.41, \qquad (4.51)$$

from which we find the error in the estimation that $x_0 = 5$ is the root of the equation

$$\Delta x = -\frac{5}{e^5} = -0.0336. \qquad (4.52)$$

Take $x = 4.9651$. In this case

$$F(4.9651) = 0.002, \qquad F'(4.9651) = -138.32, \qquad (4.53)$$

which gives $\Delta x = -0.00001$.

Take $x = 4.956$. In this case

$$F(4.956) = 1.252, \qquad F'(4.956) = -135.77, \qquad (4.54)$$

which gives $\Delta x = -0.009$.

Thus, $x = 4.9651$ is very close to the exact root of the equation.

Having the value of x at which $I(\lambda)$ is maximal, we find from $hc/\lambda k_B T = x$ the Wien displacement law

$$\lambda_{max} T = \frac{hc}{4.9651 k_B} = \text{constant}. \qquad (4.55)$$

Solution (b)

Consider the energy density distribution in terms of frequency

$$I(v) = N(v)\langle E \rangle, \qquad (4.56)$$

where $N(v) = 8\pi v^2 / c^3$ is the number of modes per unit volume and unit frequency (see Section 3.1 of the textbook for the derivation), and $\langle E \rangle$ is the average energy of a single mode.

Thus, using Planck's quantum hypothesis, the energy density distribution in terms of frequency can be written as

$$I(v) = \frac{8\pi v^2}{c^3} \frac{hv}{e^{hv/k_B T} - 1} = \frac{8\pi h}{c^3} \frac{v^3}{e^{hv/k_B T} - 1}. \qquad (4.57)$$

Substituting for $hv/k_B T = x$, we can write the energy density distribution as

$$I(v) \equiv I(x) = \frac{8\pi h}{c^3} \frac{\left(\frac{k_B T}{h} x\right)^3}{e^x - 1} = A \frac{x^3}{e^x - 1}, \qquad (4.58)$$

where $A = 8\pi (k_B T)^3 / (c^3 h^2)$.

We find the maximum of $I(x)$ by solving the equation

$$\frac{dI(x)}{dx} = 0. \qquad (4.59)$$

Thus

$$\frac{dI(x)}{dx} = A \frac{3x^2 (e^x - 1) - x^3 e^x}{(e^x - 1)^2} = A \frac{x^2 [3 (e^x - 1) - x e^x]}{(e^x - 1)^2}. \qquad (4.60)$$

Hence

$$\frac{dI(x)}{dx} = 0 \quad \text{when} \quad 3 (e^x - 1) - x e^x = 0. \qquad (4.61)$$

This equation cannot be solved exactly. Therefore, we will apply an approximate method outlined in part (a).

One can see that there are two roots of the above equation: $x = 0$ (exact root) and an approximate root $x \approx 3$ (as $e^3 \gg 1$). The root $x = 0$ is unphysical as it would correspond to $T \to \infty$, so we will focus on the root $x \approx 3$.

Let

$$F(x) = 3 (e^x - 1) - x e^x. \qquad (4.62)$$

Then

$$F'(x) = 3e^x - e^x - xe^x = (2 - x)e^x. \tag{4.63}$$

Thus, for $x \equiv x_0 = 3$:

$$F(3) = 3\left(e^3 - 1\right) - 3e^3 = -3,$$
$$F'(3) = -e^3 = -20.085, \tag{4.64}$$

from which we find the error in the estimation that $x_0 = 3$ is the root of the equation

$$\Delta x = -\frac{-3}{-e^3} = -0.15. \tag{4.65}$$

Take $x = 2.85$. In this case

$$F(2.85) = -0.41, \qquad F'(2.85) = -14.69, \tag{4.66}$$

which gives $\Delta x = -0.028$.

Take $x = 2.82$. In this case

$$F(2.82) = 0.02, \qquad F'(2.82) = -13.757, \tag{4.67}$$

which gives $\Delta x = -0.0014$.

Thus, $x = 2.82$ is a better root.

Hence, for $x = 2.82$, the corresponding frequency is

$$\nu_{max} = \frac{k_B T}{h} x = \frac{2.82 k_B}{h} T. \tag{4.68}$$

In the Tutorial Problem 4.6(a), we found that

$$\lambda_{max} = \frac{hc}{4.9651 k_B T}. \tag{4.69}$$

If we apply the relation $\lambda_{max} = c/\nu_{max}$, we obtain for ν_{max}

$$\nu_{max} = \frac{4.9651 k_B}{h} T. \tag{4.70}$$

This result differs from that in Eq. (4.68), and in this case, ν_{max} is larger than that calculated exactly.

The reason is that λ and ν are not linearly dependent quantities, $\lambda = c/\nu$, and then the densities of modes $N(\lambda)$ and $N(\nu)$ are not linear functions.

Problem 4.7

Derive the Stefan–Boltzmann law evaluating the integral in Eq. (4.15) of the textbook.
Hint: It is convenient to evaluate the integral by introducing a dimensionless variable

$$x = \frac{hc}{\lambda k_B T}. \tag{4.71}$$

Solution

To actually evaluate Eq. (4.15), it is convenient to simplify Planck's formula

$$I(\lambda) = \frac{8\pi hc}{\lambda^5 \left(e^{hc/\lambda k_B T} - 1 \right)} \tag{4.72}$$

by introducing a dimensionless variable

$$x = \frac{hc}{\lambda k_B T}. \tag{4.73}$$

When we change the variable in Planck's formula from λ to x:

$$\lambda = \frac{hc}{k_B T} \frac{1}{x} \qquad \text{so that} \qquad d\lambda = -\frac{hc}{k_B T} \frac{1}{x^2} dx, \tag{4.74}$$

we find that in terms of x, the total intensity I becomes

$$I = \frac{c}{4} \int_0^\infty I(\lambda) d\lambda = \frac{c}{4} \int_0^\infty dx \, \frac{8\pi hc}{\left(\frac{hc}{k_B T}\right)^5} \frac{x^5}{e^x - 1} \frac{hc}{k_B T x^2}$$

$$= \int_0^\infty dx \, \frac{2\pi hc^2}{\left(\frac{hc}{k_B T}\right)^4} \frac{x^3}{e^x - 1} = 2\pi hc^2 \left(\frac{k_B T}{hc}\right)^4 \int_0^\infty dx \, \frac{x^3}{e^x - 1}. \tag{4.75}$$

Since

$$\int_0^\infty \frac{x^3 dx}{e^x - 1} = \frac{\pi^4}{15}, \tag{4.76}$$

we finally obtain

$$I = 2\pi hc^2 \left(\frac{k_B T}{hc}\right)^4 \frac{\pi^4}{15} = \sigma T^4, \tag{4.77}$$

where

$$\sigma = \frac{2\pi^5 k_B^4}{15 h^3 c^2} = 5.67 \times 10^{-8} \quad [\text{W/m}^2 \cdot \text{K}^4]. \qquad (4.78)$$

The constant σ determined from experimental results agrees perfectly with the above value derived from Planck's formula.

Problem 4.10

Consider Compton scattering.

(a) Show that $\Delta E / E$, the fractional change in photon energy in the Compton effect, satisfies

$$\frac{\Delta E}{E} = \frac{h\nu'}{m_0 c^2} (1 - \cos\alpha), \qquad (4.79)$$

where ν' is the frequency of the scattered photon and $\Delta E = E - E'$.

(b) Show that the relation between the directions of motion of the scattered photon and the recoil electron in Compton scattering is

$$\cot\frac{\alpha}{2} = \left(1 + \frac{h\nu}{m_0 c^2}\right) \tan\theta, \qquad (4.80)$$

where α is the angle of the scattered photon, θ is the angle of the recoil electron, and ν is the frequency of the incident light.

Solution (a)

Use the Compton formula written in terms of the momenta of the incident (p) and scattered (p') photons

$$p - p' = \frac{p p'}{m_0 c} (1 - \cos\alpha). \qquad (4.81)$$

Since $p = E/c$, we can write the above formula in terms of the energies of the incident (E) and scattered (E') photons

$$E - E' = \frac{E E'}{m_0 c^2} (1 - \cos\alpha). \qquad (4.82)$$

Introducing a notation $\Delta E = E - E'$, and using $E' = h\nu'$, we obtain

$$\frac{\Delta E}{E} = \frac{h\nu'}{m_0 c^2} (1 - \cos\alpha). \qquad (4.83)$$

Solution (b)

As in Solution (a), consider the Compton formula in terms of the momenta of the incident and scattered photons

$$p - p' = \frac{pp'}{m_0 c}(1 - \cos\alpha). \tag{4.84}$$

We will express the momentum of the scattered photons, p', in terms of the momentum p of the incident photons using the conservation of the momentum components. Choose the geometry of the Compton scattering as shown in Fig. 4.1.

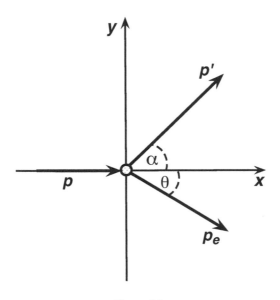

Figure 4.1

The conservation of the x and y components of the momentum leads to two equations

$$x \quad \text{component:} \quad p = p' \cos\alpha + p_e \cos\theta,$$
$$y \quad \text{component:} \quad 0 = p' \sin\alpha - p_e \sin\theta. \tag{4.85}$$

Eliminating p_e from these equations, we find that

$$p' = \frac{p \tan\theta}{\sin\alpha + \tan\theta \cos\alpha}. \tag{4.86}$$

Substituting this into Eq. (4.84), we obtain

$$p\left(1 - \frac{\tan\theta}{\sin\alpha + \tan\theta\cos\alpha}\right) = \frac{p^2}{m_0 c}\frac{\tan\theta}{(\sin\alpha + \tan\theta\cos\alpha)}(1 - \cos\alpha),$$

$$(4.87)$$

which can be written as

$$\frac{\sin\alpha}{1 - \cos\alpha} = \left(\frac{p}{m_0 c} + 1\right)\tan\theta. \qquad (4.88)$$

Since

$$\frac{\sin\alpha}{1 - \cos\alpha} = \cot\frac{\alpha}{2} \qquad \text{and} \qquad p = \frac{E}{c} = \frac{h\nu}{c}, \qquad (4.89)$$

we finally obtain

$$\cot\frac{\alpha}{2} = \left(1 + \frac{h\nu}{m_0 c^2}\right)\tan\theta. \qquad (4.90)$$

This formula shows that one can test the Compton effect by measuring the angles α and θ instead of measuring the wavelength λ' of the scattered photons.

Chapter 5

Bohr Model

Problem 5.2

Suppose that the electron is a spherical shell of radius r_e and all the electron's charge is evenly distributed on the shell. Using the formula for the energy of a charged shell, calculate the classical electron radius. Compare the size of the electron with the size of an atomic nucleus.

Solution

We know from classical electromagnetism that the energy of a charged shell is

$$E = \frac{e^2}{4\pi \varepsilon_0 r_e}. \tag{5.1}$$

Since $E = mc^2$, we find

$$r_e = \frac{e^2}{4\pi \varepsilon_0 mc^2} = 2.82 \times 10^{-15} \text{ m}. \tag{5.2}$$

This is the allowed classical electron radius. It is about the size of an atomic nucleus. The size of the electron cannot be smaller than this; otherwise, the electron's mass would be larger.

Problems and Solutions in Quantum Physics
Zbigniew Ficek
Copyright © 2016 Pan Stanford Publishing Pte. Ltd.
ISBN 978-981-4669-36-8 (Hardcover), 978-981-4669-37-5 (eBook)
www.panstanford.com

However, according to experiments, the electron is smaller, and yet its mass is not larger. Thus, classical electromagnetism must be revised for elementary particles.

Problem 5.4

Show that in the Bohr atom model, the electron's orbits in a hydrogen-like atom are quantized with the radius $r = n^2 a_0/Z$, where $a_0 = 4\pi\varepsilon_0\hbar^2/me^2$ is the Bohr radius, $n = 1, 2, \ldots$, and Z is atomic number. $Z = 1$ refers to a hydrogen atom, $Z = 2$ to a helium (He^+) ion, and so on.

Solution

From the classical equation of motion for the electron in a hydrogen-like atom (Coulomb force = centripetal force)

$$\frac{Ze^2}{4\pi\varepsilon_0 r^2} = m\frac{v^2}{r}, \tag{5.3}$$

we find the velocity of the electron

$$v = \sqrt{\frac{Ze^2}{4\pi\varepsilon_0 mr}}. \tag{5.4}$$

Bohr postulated that the angular momentum of the electron is quantized with

$$L = n\hbar, \qquad n = 1, 2, 3, \ldots \qquad \left(\hbar = \frac{h}{2\pi}\right). \tag{5.5}$$

Since

$$L = mvr = \sqrt{\frac{Zme^2 r}{4\pi\varepsilon_0}}, \tag{5.6}$$

we obtain

$$\frac{Zme^2 r}{4\pi\varepsilon_0} = n^2\hbar^2, \tag{5.7}$$

from which we find

$$r = n^2\frac{a_0}{Z}, \qquad \text{where} \qquad a_0 = \frac{4\pi\varepsilon_0\hbar^2}{me^2}. \tag{5.8}$$

Problem 5.5

The magnetic dipole moment $\vec{\mu}$ of a current loop is defined by $\vec{\mu} = I\vec{S}$, where I is the current and $\vec{S} = S\hat{n}$ is the area of the loop, with $\hat{n}$, the unit vector, normal to the plane of the loop. A current loop may be represented by a charge e rotating at constant speed in a circular orbit. Use the classical model of the orbital motion of the electron and Bohr's quantization postulate to show that the magnetic dipole moment of the loop is quantized such that

$$\mu = n\,m_{\mathrm{B}}, \qquad n = 1, 2, 3, \ldots, \tag{5.9}$$

where $m_{\mathrm{B}} = e\hbar/2m$ is the Bohr magneton, and m is the mass of the electron.

Solution

Denote the radius of the electron's orbit by r and the linear velocity of the electron by $v = \omega r$, where ω is the angular velocity. Then the period of revolution is

$$T = \frac{2\pi}{\omega} = \frac{2\pi r}{v}. \tag{5.10}$$

Hence, the current induced by the revolting electron is

$$I = \frac{e}{T} = \frac{ev}{2\pi r}. \tag{5.11}$$

We know from electromagnetism that current produces a magnetic field and a current loop closing some area creates a magnetic moment. The magnetic moment is equal to the product of the area of the plane loop and the magnitude of the circulating current:

$$\vec{\mu} = I\vec{S} = IS\hat{n}, \tag{5.12}$$

where $S = \pi r^2$ is the area closed by the loop (the orbit of the revolting electron), $\hat{n}$ is the unit vector perpendicular to the plane of the loop and oriented along the direction set by the right-hand rule.

Thus

$$\vec{\mu} = \frac{ev}{2\pi r}\pi r^2 \hat{n} = \frac{1}{2}evr\hat{n}. \tag{5.13}$$

From the definition of the angular momentum

$$\vec{L} = \vec{p} \times \vec{r} = mvr\hat{n}, \tag{5.14}$$

where $\vec{p} = m\vec{v}$, we find that

$$\vec{\mu} = \frac{1}{2}evr\hat{n} = \frac{e}{2m}\vec{L}. \tag{5.15}$$

Since

$$L = n\hbar, \tag{5.16}$$

we find that

$$\vec{\mu} = n\frac{e\hbar}{2m}\hat{n} = n\, m_B\hat{n}, \tag{5.17}$$

where $m_B = e\hbar/2m = 9.27 \times 10^{-24}$ [A·m^2] is the Bohr magneton.

Problem 5.6

Consider an experiment. A student is at a distance of 10 m from a light source whose power is $P = 40$ W.

(a) How many photons strike the student's eye if the wavelength of light is 589 nm (yellow light) and the radius of the pupil (a variable aperture through which light enters the eye) is 2 mm.
(b) At what distance from the source, only one photon would strike the student's eye.

Solution (a)

The intensity of light at a distance of 10 m from the source is

$$I = \frac{P}{4\pi r^2} = \frac{40}{4\pi (10)^2} = 0.032 \quad \left[\frac{W}{m^2}\right]. \tag{5.18}$$

Energy of a single photon of wavelength $\lambda = 589$ nm is

$$E = h\nu = \frac{hc}{\lambda} = \frac{6.63 \times 10^{-34} \times 3 \times 10^8}{589 \times 10^{-9}} = 0.034 \times 10^{-17} \text{ [J]}. \tag{5.19}$$

The rate at which energy is absorbed by the eye is given by

$$R = IA = 0.032 \times \pi \times (2 \times 10^{-3})^2 = 402.1 \times 10^{-9} \quad \left[\frac{J}{s}\right], \tag{5.20}$$

where A is the area of the pupil.

Hence, we find that the number of photons striking the eye per second is given by

$$n = \frac{R}{E} = \frac{402.1 \times 10^{-9}}{0.034 \times 10^{-17}} = 11{,}826.5 \times 10^{8} \approx 12 \times 10^{11} \left[\frac{\text{photons}}{\text{s}} \right].$$

$$(5.21)$$

Solution (b)

We have to find the distance at which the rate of absorption of light per second is equal to the energy of a single photon, i.e.,

$$R = I A = E. \qquad (5.22)$$

Since $I = P/(4\pi r^2)$, we have

$$\frac{P A}{4\pi r^2} = E, \qquad (5.23)$$

from which we find

$$r^2 = \frac{P A}{4\pi E}. \qquad (5.24)$$

Hence

$$r = \sqrt{\frac{P A}{4\pi E}} = \sqrt{\frac{40 \times \pi \times (2 \times 10^{-3})^2}{4\pi \times 0.034 \times 10^{-17}}}$$

$$= \sqrt{118 \times 10^{12}} \approx 11 \times 10^{6} \text{ [m]} = 11 \times 10^{3} \text{ [km]}. \qquad (5.25)$$

Chapter 6

Duality of Light and Matter

Problem 6.3

Determine where a particle is most likely to be found whose wave function is given by

$$\Psi(x) = \frac{1 + ix}{1 + ix^2}. \tag{6.1}$$

Solution

The probability density of finding the particle at a point x is given by

$$|\Psi(x)|^2 = \Psi(x)\,\Psi^*(x) = \frac{1 + ix}{1 + ix^2}\frac{1 - ix}{1 - ix^2} = \frac{1 + x^2}{1 + x^4}. \tag{6.2}$$

The particle is most likely to be found at points for which $d|\Psi|^2/dx = 0$. Since

$$\frac{d\,|\Psi|^2}{dx} = \frac{2x(1 + x^4) - 4x^3(1 + x^2)}{(1 + x^4)^2}, \tag{6.3}$$

we find that $d|\Psi|^2/dx = 0$ when

$$2x(1 + x^4) - 4x^3(1 + x^2) = 0. \tag{6.4}$$

Problems and Solutions in Quantum Physics
Zbigniew Ficek
Copyright © 2016 Pan Stanford Publishing Pte. Ltd.
ISBN 978-981-4669-36-8 (Hardcover), 978-981-4669-37-5 (eBook)
www.panstanford.com

This equation can be simplified to

$$x^4 + 2x^2 - 1 = 0, \tag{6.5}$$

which, after substituting $x^2 = z$, reduces to a quadratic equation

$$z^2 + 2z - 1 = 0, \tag{6.6}$$

whose the roots are

$$z_1 = -1 + \sqrt{2} \qquad \text{and} \qquad z_2 = -1 - \sqrt{2}. \tag{6.7}$$

Thus $d|\Psi|^2/dx = 0$ when

$$x_1^2 = -1 + \sqrt{2} \qquad \text{and} \qquad x_2^2 = -1 - \sqrt{2}. \tag{6.8}$$

Since $x^2 > 0$, the only solution we can accept is

$$x_1 = \pm\sqrt{-1 + \sqrt{2}}. \tag{6.9}$$

Problem 6.4

The wave function of a free particle at $t = 0$ is given by

$$\Psi(x, 0) = \begin{cases} 0 & x < -b, \\ A & -b \leq x \leq 3b, \\ 0 & x > 3b. \end{cases} \tag{6.10}$$

(a) Using the fact that the probability is normalized to one, i.e.,

$$\int_{-\infty}^{+\infty} |\Psi(x, 0)|^2 dx = 1, \tag{6.11}$$

find the constant A. (You can assume that A is real.)

(b) What is the probability of finding the particle within the interval $x \in [0, b]$ at time $t = 0$?

Solution (a)

The constant A is found from the normalization condition, which can be written as

$$1 = \int_{-\infty}^{+\infty} |\Psi|^2 dx = \int_{-\infty}^{-b} |\Psi|^2 dx + \int_{-b}^{3b} |\Psi|^2 dx + \int_{3b}^{+\infty} |\Psi|^2 dx$$

$$= 0 + \int_{-b}^{3b} |\Psi|^2 dx + 0 = A^2 \int_{-b}^{3b} dx = 4bA^2. \tag{6.12}$$

Hence

$$A = \frac{1}{2\sqrt{b}}.$$

(6.13)

Solution (b)

The probability of finding the particle within the interval $x \in [0, b]$ at time $t = 0$ is given by

$$\int_0^b |\Psi|^2 dx = A^2 \int_0^b dx = \frac{1}{4b} b = \frac{1}{4}.$$

(6.14)

Problem 6.5

The state of a free particle at $t = 0$ confined between two walls separated by a is described by the following wave function:

$$\Psi(x, 0) = \Psi_{max} \sin\left(\frac{n\pi}{a}x\right), \qquad 0 \leq x \leq a,$$

$$\Psi(x, 0) = 0, \qquad x > a, \quad \text{and} \quad x < 0.$$

(6.15)

(a) Find the amplitude Ψ_{max} using the normalization condition.
(b) What is the probability density of finding the particle at $x = 0, a/2$, and a. How does the result depend on n?
(c) Calculate the probability of finding the particle in the regions $\frac{a}{2} \leq x \leq a$ and $\frac{3a}{4} \leq x \leq a$, for $n = 1$ and $n = 2$.

Solution (a)

From the normalization condition, we find

$$1 = \int_{-\infty}^{+\infty} |\Psi|^2 dx = \int_{-\infty}^{0} |\Psi|^2 dx + \int_0^a |\Psi|^2 dx + \int_a^{+\infty} |\Psi|^2 dx$$

$$= 0 + \int_0^a |\Psi|^2 dx + 0 = |\Psi_{max}|^2 \int_0^a \sin^2\left(\frac{n\pi}{a}x\right) dx$$

$$= \frac{1}{2}|\Psi_{max}|^2 \int_0^a \left[1 - \cos\left(\frac{2n\pi}{a}x\right)\right] dx$$

$$= \frac{1}{2}|\Psi_{max}|^2 \left[x - \frac{a}{2n\pi}\sin\left(\frac{2n\pi}{a}x\right)\right]_0^a = \frac{a}{2}|\Psi_{max}|^2,$$

(6.16)

as $\sin(2n\pi) = \sin(0) = 0$. Hence

$$|\Psi_{\max}| = \sqrt{\frac{2}{a}}. \tag{6.17}$$

Solution (b)

From the definition of the probability density, $P_d = |\Psi(x)|^2$, we find

$$P_d = \frac{2}{a}\sin^2\left(\frac{n\pi}{a}x\right). \tag{6.18}$$

Thus, at $x = 0$, the probability density $P_d = 0$ is independent of n.

Similarly, at $x = a$, the probability density $P_d = 0$ is independent of n.

At $x = a/2$

$$P_d = \frac{2}{a}\sin^2\left(\frac{n\pi}{a}\right). \tag{6.19}$$

Hence, for odd n ($n = 1, 3, 5, \ldots$), the probability density is maximum (equal to $2/a$), whereas for even n ($n = 2, 4, 6, \ldots$), the probability density $P_d = 0$.

Solution (c)

The probability of finding the particle in the region $\frac{a}{2} \leq x \leq a$ is given by

$$
\begin{aligned}
P &= \int_{a/2}^{a} |\Psi|^2 dx = |\Psi_{\max}|^2 \int_{a/2}^{a} \sin^2\left(\frac{n\pi}{a}x\right) dx \\
&= \frac{1}{2}|\Psi_{\max}|^2 \int_{a/2}^{a} \left[1 - \cos\left(\frac{2n\pi}{a}x\right)\right] dx \\
&= \frac{1}{a}\left[x - \frac{a}{2n\pi}\sin\left(\frac{2n\pi}{a}x\right)\right]_{a/2}^{a} = \frac{1}{2},
\end{aligned} \tag{6.20}
$$

for both $n = 1$ and $n = 2$, as $\sin(2n\pi) = \sin(n\pi) = 0$ for all n.

Similarly as above, we find that the probability of finding the particle in the region $\frac{3a}{4} \leq x \leq a$ is given by

$$
\begin{aligned}
P &= \int_{3a/4}^{a} |\Psi|^2 dx = |\Psi_{max}|^2 \int_{3a/4}^{a} \sin^2 \left(\frac{n\pi}{a} x \right) dx \\
&= \frac{1}{2} |\Psi_{max}|^2 \int_{3a/4}^{a} \left[1 - \cos \left(\frac{2n\pi}{a} x \right) \right] dx \\
&= \frac{1}{a} \left[x - \frac{a}{2n\pi} \sin \left(\frac{2n\pi}{a} x \right) \right]_{3a/4}^{a} = \frac{1}{4} + \frac{1}{2n\pi} \sin \left(\frac{3}{2} n\pi \right).
\end{aligned}
$$
$$(6.21)$$

Now since $\sin \left(\frac{3}{2} n\pi \right) = -1$ for $n = 1$, and $\sin \left(\frac{3}{2} n\pi \right) = 0$ for $n = 2$, we have the result

$$
P = \frac{1}{4} \left(1 - \frac{2}{\pi} \right) \qquad \text{for} \qquad n = 1,
$$

$$
P = \frac{1}{4} \qquad \text{for} \qquad n = 2. \qquad\qquad (6.22)
$$

Problem 6.6

The time-independent wave function of a particle is given by

$$
\Psi(x) = Ae^{-|x|/\sigma}, \qquad\qquad (6.23)
$$

where A and σ are constants.

(a) Sketch this function and find A in terms of σ such that $\Psi(x)$ is normalized.
(b) Find the probability that the particle will be found in the region $-\sigma \leq x \leq \sigma$.

Solution (a)

The wave function $\Psi(x)$ can be written as

$$
\Psi(x) = \begin{cases} Ae^{x/\sigma} & \text{for} & x < 0, \\ \\ Ae^{-x/\sigma} & \text{for} & x \geq 0. \end{cases} \qquad\qquad (6.24)
$$

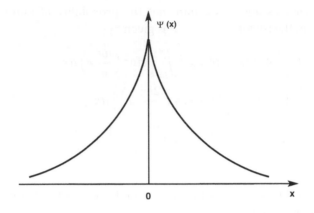

$\Psi(x)$

0

x

Figure 6.1

The wave function is symmetric, decaying exponentially from the origin in both directions, as illustrated in Fig. 6.1.

From the normalization condition

$$\int_{-\infty}^{+\infty} |\Psi(x)|^2 dx = 1, \qquad (6.25)$$

we have

$$\int_{-\infty}^{+\infty} |\Psi(x)|^2 dx = |A|^2 \int_{-\infty}^{+\infty} e^{-2|x|/\sigma} dx$$

$$= |A|^2 \left\{ \int_{-\infty}^{0} e^{2x/\sigma} dx + \int_{0}^{+\infty} e^{-2x/\sigma} dx \right\}. \qquad (6.26)$$

We can change the variable x into $-x$ in the first integral and obtain

$$\int_{-\infty}^{0} e^{2x/\sigma} dx = -\int_{+\infty}^{0} e^{-2x/\sigma} dx = \int_{0}^{+\infty} e^{-2x/\sigma} dx. \qquad (6.27)$$

Hence

$$1 = \int_{-\infty}^{+\infty} |\Psi(x)|^2 dx = 2|A|^2 \int_{0}^{+\infty} e^{-2x/\sigma} dx$$

$$= 2|A|^2 \left(-\frac{\sigma}{2}\right) e^{-2x/\sigma} \Big|_{0}^{+\infty} = \sigma |A|^2. \qquad (6.28)$$

Thus

$$A = \sqrt{\frac{1}{\sigma}}. \qquad (6.29)$$

Solution (b)

The probability of finding the particle in the region $-\sigma \leq x \leq \sigma$ is

$$P = \int_{-\sigma}^{\sigma} |\Psi(x)|^2 dx = |A|^2 \int_{-\sigma}^{\sigma} e^{-2|x|/\sigma} dx$$

$$= |A|^2 \left\{ \int_{-\sigma}^{0} e^{2x/\sigma} dx + \int_{0}^{\sigma} e^{-2x/\sigma} dx \right\} = 2|A|^2 \int_{0}^{\sigma} e^{-2x/\sigma} dx$$

$$= 2|A|^2 \left(-\frac{\sigma}{2} \right) e^{-2x/\sigma} \Big|_{0}^{\sigma} = -\left(e^{-2} - 1 \right) = 1 - e^{-2} = 0.856.$$

$$(6.30)$$

Thus, there is about a 86% chance that the particle will be found in the region $-\sigma \leq x \leq \sigma$.

Problem 6.7

We have calculated the phase velocity u using the relativistic formula for energy. Calculate the phase velocity for the non-relativistic case. Does the relativistic result for u tends to the corresponding non-relativistic result as the velocity of the particle becomes small compared to the speed of light?

Solution

In the non-relativistic case, the energy of the particle is given by

$$E = \frac{p^2}{2m}, \qquad (6.31)$$

where p is the momentum of the particle.

Since $p = \hbar k$ and $E = \hbar \omega$, we have

$$E = \frac{p^2}{2m} = \frac{\hbar^2}{2m} k^2 = \hbar \omega. \qquad (6.32)$$

Thus, in the non-relativistic case

$$\omega = \frac{\hbar}{2m} k^2. \qquad (6.33)$$

With this relation between ω and k, we find that the phase velocity is

$$u = \frac{\omega}{k} = \frac{\hbar}{2m}k, \tag{6.34}$$

and the group velocity is

$$v_g = \frac{d\omega}{dk} = \frac{\hbar}{m}k = \frac{p}{m} = v. \tag{6.35}$$

Therefore,

$$u = \frac{1}{2}v_g = \frac{1}{2}v. \tag{6.36}$$

In the relativistic case

$$u = \frac{c^2}{v_g} = \frac{c^2}{v}. \tag{6.37}$$

Thus, the relativistic case does not tend to the non-relativistic case when $v \ll c$. Normally, a relativistic result in physics tends to the corresponding non-relativistic result as the velocity involved becomes small compared to the speed of light. This is clearly not the case for the above two expressions for phase velocity. The reason is that the expression for the relativistic energy

$$E^2 = p^2c^2 + (m_0c^2)^2 \tag{6.38}$$

includes the rest-mass term, m_0c^2, whereas the expression for the non-relativistic energy $E = p^2/2m$ does not include the rest-mass term.

Problem 6.8

We know that the group velocity v_g of the wave packet of a particle of mass m is equal to the velocity v of the particle. Show that the total energy of the particle is $E = \hbar\omega$, the same which holds for photons.

Solution

From the definition of momentum

$$\vec{p} = m\vec{v}, \tag{6.39}$$

and the fact that the velocity of the particle $\vec{v} = \vec{v}_g$ and $\vec{p} = \hbar\vec{k}$, we have

$$m\vec{v}_g = \hbar\vec{k}. \tag{6.40}$$

Using the definition of the group velocity, which in three dimensions can be written as

$$\vec{v}_g = \nabla_k\omega, \tag{6.41}$$

where

$$\nabla_k\omega = \frac{\partial\omega}{\partial k_x}\vec{i} + \frac{\partial\omega}{\partial k_y}\vec{j} + \frac{\partial\omega}{\partial k_z}\vec{k} \tag{6.42}$$

is the gradient over the components of $\vec{k}$ (k_x, k_y, k_z), we have

$$m\nabla_k\omega = \hbar\vec{k}. \tag{6.43}$$

Integrating this equation over k, we obtain

$$m\omega = \frac{\hbar}{2}\left(k_x^2 + k_y^2 + k_z^2\right) + C, \tag{6.44}$$

where C is a constant.

Hence, multiplying both sides by $\hbar$ and dividing by m, we obtain

$$\hbar\omega = \frac{\hbar^2}{2m}k^2 + A, \tag{6.45}$$

where $A = \hbar C/m$ is a constant.

Since $\hbar^2 k^2 = p^2$, we see that the right-hand side of the above equation is the total energy E of the particle. Thus,

$$\hbar\omega = E, \tag{6.46}$$

which is the same that holds for photons.

Problem 6.11

The time required for a wave packet to move the distance equal to the width of the wave packet is $\Delta t = \Delta x/v_g$, where Δx is the width of the wave packet. Show that the time Δt and the uncertainty in the energy of the particle satisfy the uncertainty relation

$$\Delta E\,\Delta t = h, \tag{6.47}$$

where $\Delta E = \hbar\Delta\omega$.

Solution

Since

$$\Delta x = v_{\mathrm{g}} \Delta t = \frac{\Delta \omega}{\Delta k} \Delta t, \tag{6.48}$$

we find that the uncertainty relation

$$\Delta x \Delta k = 2\pi, \tag{6.49}$$

can be written as

$$\Delta x \Delta k = \frac{\Delta \omega}{\Delta k} \Delta t \Delta k = \Delta \omega \Delta t = 2\pi. \tag{6.50}$$

Multiplying both sides of the above equation by $\hbar$, we obtain

$$\hbar \Delta \omega \Delta t = 2\pi \hbar = h. \tag{6.51}$$

Since $\Delta E = \hbar \Delta \omega$, we finally obtain the energy and time uncertainty relation

$$\Delta E \Delta t = h. \tag{6.52}$$

In the above relation, ΔE is the uncertainty in our knowledge of the energy E of a system and Δt is the time interval characteristic of the rate of changes in the system's energy.

Problem 6.12

The amplitude $A(k)$ of the wave function

$$\Psi(x, t) = \int_{-\infty}^{+\infty} A(k) e^{i(kx - \omega_k t)} dk \tag{6.53}$$

is given by

$$A(k) = \begin{cases} 1 & \text{for} \quad k_0 - \frac{1}{2}\Delta k \leq k \leq k_0 + \frac{1}{2}\Delta k, \\ 0 & \text{for} \quad k > k_0 + \frac{1}{2}\Delta k \quad \text{and} \quad k < k_0 - \frac{1}{2}\Delta k. \end{cases} \tag{6.54}$$

(a) Show that the wave function can be written as

$$\Psi(x, t) = \frac{\sin z}{z} \Delta k \, e^{i(k_0 x - \omega_0 t)}, \tag{6.55}$$

where $z = \frac{1}{2}\Delta k(x - v_{\mathrm{g}} t)$.

(b) Sketch the function $f(z) = \sin z / z$ and find the width of the main maximum of $f(z)$.

(*Hint:* For $f(z)$, one might define a suitable width as the spacing between its first two zeros.)

Solution (a)

With the shape of the amplitude $A(k)$:

$$A(k) = \begin{cases} 1 & \text{for} \quad k_0 - \frac{1}{2}\Delta k \le k \le k_0 + \frac{1}{2}\Delta k, \\ 0 & \text{for} \quad k > k_0 + \frac{1}{2}\Delta k, \quad \text{and} \quad k < k_0 - \frac{1}{2}\Delta k, \end{cases} \quad (6.56)$$

the wave packet has the form

$$\Psi(x, t) = \int_k A(k) e^{i(kx - \omega_k t)} dk = \int_{k_0 - \frac{1}{2}\Delta k}^{k_0 + \frac{1}{2}\Delta k} e^{i(kx - \omega_k t)} dk. \quad (6.57)$$

Taking $k = k_0 + \beta$, and expanding ω_k into a Taylor series about $k = k_0$, we get

$$\omega_k = \omega_{k_0 + \beta} = \omega_0 + \left(\frac{d\omega}{d\beta}\right)_{k_0} \beta + \frac{1}{2}\left(\frac{d^2\omega}{d\beta^2}\right)_{k_0} \beta^2 + \dots, \quad (6.58)$$

where $\omega_0 = \omega_{k_0}$.

If we take only the first two terms of the series and substitute to $\Psi(x, t)$, we obtain

$$\Psi(x, t) = e^{i(k_0 x - \omega_0 t)} \int_{-\frac{1}{2}\Delta k}^{\frac{1}{2}\Delta k} d\beta e^{i\beta(x - v_g t)}, \quad (6.59)$$

where $v_g = \left(\frac{d\omega}{d\beta}\right)_{k_0}$ is the group velocity of the packet.

Performing the integration, we obtain

$$\Psi(x, t) = e^{i(k_0 x - \omega_0 t)} \left. \frac{e^{i\beta(x - v_g t)}}{i(x - v_g t)} \right|_{-\frac{1}{2}\Delta k}^{\frac{1}{2}\Delta k}$$

$$= e^{i(k_0 x - \omega_0 t)} \frac{\left[e^{i(x - v_g t)\frac{1}{2}\Delta k} - e^{-i(x - v_g t)\frac{1}{2}\Delta k} \right]}{i(x - v_g t)}$$

$$= \frac{2e^{i(k_0 x - \omega_0 t)}}{(x - v_g t)} \sin\left[\frac{1}{2}\Delta k(x - v_g t) \right] = \frac{\sin z}{z}\Delta k \, e^{i(k_0 x - \omega_0 t)}, \quad (6.60)$$

where $z = \frac{1}{2}\Delta k(x - v_g t)$.

Solution (b)

Figure 6.2 shows the variation of $f(z) = \sin z/z$ with z. The width of the main maximum can be approximated by the distance between the first two zeros of the function $f(z)$. It is seen from Fig. 6.2 that the first zeros are at $z = \pm\pi$. Thus, the width of the main maximum is 2π.

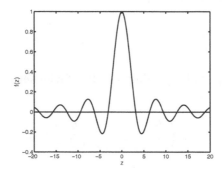

Figure 6.2 Variation of $f(z) = \sin z/z$ with z.

Problem 6.13

Calculate $A(k)$, the inverse Fourier transform

$$A(k) = \frac{1}{\sqrt{2\pi}} \int_{-\infty}^{+\infty} \Psi(x, 0)e^{-ikx} dx \qquad (6.61)$$

of the triangular wave packet

$$\Psi(x, 0) = \begin{cases} 1 + \frac{x}{b} & -b \le x \le 0, \\ 1 - \frac{x}{b} & 0 < x < b, \\ 0 & \text{elsewhere.} \end{cases} \qquad (6.62)$$

Draw qualitative graphs of $A(k)$ and $\Psi(x, 0)$. Next to each graph, write down its approximate "width".

Solution

With the wave packet of the form

$$\Psi(x, 0) = \begin{cases} 1 + \frac{x}{b} & -b \le x \le 0, \\ 1 - \frac{x}{b} & 0 < x < b, \\ 0 & \text{elsewhere,} \end{cases} \tag{6.63}$$

the amplitude $A(k)$ takes the form

$$A(k) = \frac{1}{\sqrt{2\pi}} \int_{-\infty}^{+\infty} \Psi(x, 0) e^{-ikx} dx$$

$$= \frac{1}{\sqrt{2\pi}} \int_{-b}^{0} \left(1 + \frac{x}{b}\right) e^{-ikx} dx + \frac{1}{\sqrt{2\pi}} \int_{0}^{b} \left(1 - \frac{x}{b}\right) e^{-ikx} dx.$$

$$\tag{6.64}$$

We can change the variable x to $-x$ in the first integral and obtain

$$A(k) = \frac{1}{\sqrt{2\pi}} \int_{0}^{b} \left(1 - \frac{x}{b}\right) \left(e^{ikx} + e^{-ikx}\right) dx$$

$$= \frac{2}{\sqrt{2\pi}} \int_{0}^{b} \left(1 - \frac{x}{b}\right) \cos(kx) dx. \tag{6.65}$$

Performing the integration, we get

$$A(k) = \frac{2}{\sqrt{2\pi}} \frac{1}{k^2 b} [1 - \cos(kb)], \tag{6.66}$$

which can be simplified to

$$A(k) = \frac{2}{\sqrt{2\pi}} \frac{1}{k^2 b} [1 - \cos(kb)] = \frac{4}{\sqrt{2\pi}} \frac{1}{k^2 b} \sin^2\left(\frac{1}{2} kb\right)$$

$$= \frac{1}{\sqrt{2\pi}} \frac{b}{\frac{1}{4} k^2 b^2} \sin^2\left(\frac{1}{2} kb\right) = \frac{b}{\sqrt{2\pi}} \frac{\sin^2\left(\frac{1}{2} kb\right)}{\left(\frac{1}{2} kb\right)^2}$$

$$= \frac{b}{\sqrt{2\pi}} \left[\frac{\sin\left(\frac{1}{2} kb\right)}{\frac{1}{2} kb}\right]^2. \tag{6.67}$$

Figure 6.3 shows the wave packet $\Psi(x, 0)$ and the amplitude $A(k)$ for $b = 1$. The width of the wave packet is $2b$, whereas the width of the amplitude $A(k)$ is 2π.

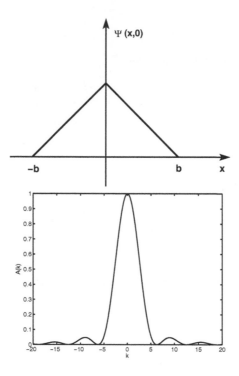

Figure 6.3

Problem 6.14

The wave function of a particle is given by a wave packet

$$\Psi(x, t) = \int_{-\infty}^{+\infty} A(k) e^{i(kx - \omega_k t)} dk. \tag{6.68}$$

Assuming that the amplitude $A(k) = \exp(-\alpha|k|)$, show that the wave function is in the form of a Lorentzian

$$\Psi(x, t) = \frac{2\alpha}{\alpha^2 + (x - v_g t)^2}. \tag{6.69}$$

(*Hint:* Expand k and ω_k in a Taylor series around $k_0 = \omega_0 = 0$.)

Solution

Since $A(k) = \exp(-\alpha|k|)$, the amplitude of the wave packet has the explicit form

$$A(k) = \begin{cases} e^{\alpha k} & \text{for} \quad k < 0, \\ e^{-\alpha k} & \text{for} \quad k \geq 0. \end{cases} \tag{6.70}$$

Therefore, the wave function can be written as

$$\Psi(x, t) = \int_{-\infty}^{+\infty} e^{-\alpha|k|} e^{i(kx - \omega_k t)} dk$$

$$= \int_{-\infty}^{0} e^{\alpha k} e^{i(kx - \omega_k t)} dk + \int_{0}^{+\infty} e^{-\alpha k} e^{i(kx - \omega_k t)} dk. \tag{6.71}$$

Since ω_k depends on k and the explicit dependence is unknown, we may expand k and ω_k in a Taylor series around $k_0 = \omega_0 = 0$, i.e., we can write

$$k \approx k_0 + \beta = \beta,$$

$$\omega_k \approx \omega_0 + \frac{d\omega_k}{dk}\beta = v_g\beta, \tag{6.72}$$

and obtain

$$\Psi(x, t) = \int_{-\infty}^{0} e^{\alpha\beta} e^{i(x - v_g t)\beta} d\beta + \int_{0}^{+\infty} e^{-\alpha\beta} e^{i(x - v_g t)\beta} d\beta . \tag{6.73}$$

We can change the variable β to $-\beta$ in the first integral and then obtain

$$\Psi(x, t) = \int_{0}^{+\infty} e^{-\alpha\beta} e^{-i(x - v_g t)\beta} d\beta + \int_{0}^{+\infty} e^{-\alpha\beta} e^{i(x - v_g t)\beta} d\beta$$

$$= \int_{0}^{+\infty} e^{-\alpha\beta} \left[e^{-i(x - v_g t)\beta} + e^{i(x - v_g t)\beta} \right] d\beta. \tag{6.74}$$

Using Euler's formula ($e^{\pm ix} = \cos x \pm i \sin x$) and performing the integration, the above wave function simplifies to

$$\Psi(x, t) = 2 \int_{0}^{+\infty} e^{-\alpha\beta} \cos \left[(x - v_g t)\beta \right] d\beta$$

$$= \frac{2e^{-\alpha\beta}}{\alpha^2 + (x - v_g t)^2}$$

$$\times \left\{ -\alpha \cos \left[(x - v_g t)\beta \right] + (x - v_g t) \sin \left[(x - v_g t)\beta \right] \right\} \Big|_{0}^{+\infty}$$

$$= \frac{2\alpha}{\alpha^2 + (x - v_g t)^2}. \tag{6.75}$$

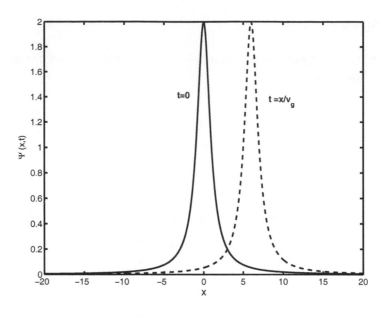

Figure 6.4

Thus, the wave packet has a Lorentzian shape. The Lorentzian is centered on $x = v_{\mathrm{g}}t$ and the width is equal to α, as seen in Fig. 6.4. Hence, if at $t = 0$ the wave packet was at $x = 0$, in time t it will move a distance $x = v_{\mathrm{g}}t$.

Chapter 7

Non-Relativistic Schrödinger Equation

Problem 7.1

Usually, we find the wave function by knowing the potential $V(x)$. Consider, however, an inverse problem where we know the wave function and would like to determine the potential that leads to the behavior described by the wave function.

Assume that a particle is confined within the region $0 \le x \le a$, and its wave function is

$$\phi(x) = \sin\left(\frac{\pi x}{a}\right). \tag{7.1}$$

Using the stationary Schrödinger equation, find the potential $V(x)$ confining the particle.

Solution

The Schrödinger equation involves the second-order derivative of the wave function. Thus, finding the second-order derivatives of the wave function

$$\frac{d\phi(x)}{dx} = \frac{\pi}{a}\cos\left(\frac{\pi x}{a}\right),$$

Problems and Solutions in Quantum Physics
Zbigniew Ficek
Copyright © 2016 Pan Stanford Publishing Pte. Ltd.
ISBN 978-981-4669-36-8 (Hardcover), 978-981-4669-37-5 (eBook)
www.panstanford.com

$$\frac{d^2\phi(x)}{dx^2} = -\left(\frac{\pi}{a}\right)^2 \sin\left(\frac{\pi x}{a}\right), \qquad (7.2)$$

the stationary Schrödinger equation then takes the form

$$\frac{\hbar^2}{2m}\left(\frac{\pi}{a}\right)^2 \sin\left(\frac{\pi x}{a}\right) + V(x)\sin\left(\frac{\pi x}{a}\right) = E\sin\left(\frac{\pi x}{a}\right). \qquad (7.3)$$

We can write this expression as

$$\left[\frac{\hbar^2}{2m}\left(\frac{\pi}{a}\right)^2 + V(x) - E\right]\sin\left(\frac{\pi x}{a}\right) = 0. \qquad (7.4)$$

This equation must be satisfied for all x within the region $0 \le x \le a$, which means that the expression in the squared brackets must be zero, i.e.,

$$\frac{\hbar^2}{2m}\left(\frac{\pi}{a}\right)^2 + V(x) - E = 0. \qquad (7.5)$$

Note that the wave function is in the form of a sine function $\sin(kx)$, which means that

$$k = \frac{\pi}{a}, \qquad (7.6)$$

so then

$$E = \frac{\hbar^2 k^2}{2m} = \frac{\hbar^2 \pi^2}{2ma^2}. \qquad (7.7)$$

Substituting this expression for E into Eq. (7.5), we easily find that $V(x) = 0$. Thus, $\phi(x)$ is the wave function of a particle moving in the potential $V(x) = 0$.

Problem 7.2

Another example of the inverse problem where we know the wave function and would like to determine the potential that leads to the behavior described by the wave function.

Consider the one-dimensional stationary wave function

$$\phi(x) = A\left(\frac{x}{x_0}\right)^n e^{-x/x_0}, \qquad (7.8)$$

where A, x_0, and n are constants.

Using the stationary Schrödinger equation, find the potential $V(x)$ and the energy E for which this wave function is an eigenfunction.

Assume that $V(x) \to 0$ as $x \to \infty$.

Solution

Consider a stationary one-dimensional Schrödinger equation

$$\left(-\frac{\hbar^2}{2m}\frac{d^2}{dx^2} + V(x)\right)\phi(x) = E\phi(x), \qquad (7.9)$$

which is the eigenvalue equation for the Hamiltonian of a particle of mass m moving in the potential $V(x)$.

Note that the equation involves the second-order derivative of the wave function. Thus, we take derivatives of the wave function

$$\frac{d\phi(x)}{dx} = A\frac{n}{x_0}\left(\frac{x}{x_0}\right)^{n-1} e^{-x/x_0} + A\left(\frac{x}{x_0}\right)^n\left(\frac{-1}{x_0}\right) e^{-x/x_0}, \quad (7.10)$$

$$\frac{d^2\phi(x)}{dx^2} = A\frac{n(n-1)}{x_0^2}\left(\frac{x}{x_0}\right)^{n-2} e^{-x/x_0}$$

$$-2A\frac{n}{x_0^2}\left(\frac{x}{x_0}\right)^{n-1} e^{-x/x_0} + A\frac{1}{x_0^2}\left(\frac{x}{x_0}\right)^n e^{-x/x_0}$$

$$= \left[\frac{n(n-1)}{x^2} - 2\frac{n}{xx_0} + \frac{1}{x_0^2}\right]\phi(x). \qquad (7.11)$$

Substituting the above result into the Schrödinger equation, we find that $\phi(x)$ is an eigenfunction with the eigenvalue E when

$$-\frac{\hbar^2}{2m}\left[\frac{n(n-1)}{x^2} - 2\frac{n}{xx_0} + \frac{1}{x_0^2}\right] = E - V(x). \qquad (7.12)$$

As $V(x) \to 0$ when $x \to \infty$, we have

$$E = -\frac{\hbar^2}{2mx_0^2}, \qquad (7.13)$$

and hence

$$V(x) = \frac{\hbar^2}{2m}\left[\frac{n(n-1)}{x^2} - \frac{2n}{xx_0}\right]. \qquad (7.14)$$

A comment: The above potential is an example of an effective potential for a hydrogen-like atom

$$V(x) = \frac{e^2}{r} - \frac{l(l+1)\hbar^2}{2mr^2}, \qquad (7.15)$$

where the first term on the right-hand side is the Coulomb potential and the second term is the so-called screening potential.

Problem 7.3

Consider the three-dimensional time-dependent Schrödinger equation of a particle of mass m moving with a potential $\hat{V}(\vec{r}, t)$:

$$i\hbar\frac{\partial\Psi(\vec{r}, t)}{\partial t} = \left(-\frac{\hbar^2}{2m}\nabla^2 + \hat{V}(\vec{r}, t)\right)\Psi(\vec{r}, t). \qquad (7.16)$$

(a) Explain, what must be assumed about the form of the potential energy to make the equation separable into a time-independent Schrödinger equation and an equation for the time dependence of the wave function.

(b) Using the condition stated in part (a), separate the time-dependent Schrödinger equation into a time-independent Schrödinger equation and an equation for the time-dependent part of the wave function.

(c) Solve the equation for the time-dependent part of the wave function and explain why the wave function of the separable Schrödinger equation is a stationary state of the particle.

Solution (a)

The Hamiltonian of the particle involved in the Schrödinger equation

$$\hat{H}(\vec{r}, t) = -\frac{\hbar^2}{2m}\nabla^2 + \hat{V}(\vec{r}, t), \qquad (7.17)$$

depends on the spatial variables through the kinetic and the potential energies, and also on time but only through the potential energy. If the potential energy is independent of time, then the Hamiltonian depends solely on the spatial variables. In other words, the Hamiltonian does not affect the time dependence of the wave function of the particle. Therefore, the wave function $\Psi(\vec{r}, t)$ can be written as a product of two parts, $\Psi(\vec{r}, t) = \phi(\vec{r})f(t)$, where $\phi(\vec{r})$ is a part of the wave function that depends solely on the spatial variables and $f(t)$ is a part that depends solely on time.

Solution (b)

If $\Psi(\vec{r}, t) = \phi(\vec{r}) f(t)$, then the Schrödinger equation takes the form

$$i\hbar\phi(\vec{r})\frac{\partial f(t)}{\partial t} = f(t)\left[-\frac{\hbar^2}{2m}\nabla^2 + \hat{V}(\vec{r})\right]\phi(\vec{r}), \qquad (7.18)$$

where we have used the fact that $\phi(\vec{r})$ is a constant for the differentiation over time and $f(t)$ is a constant for the differentiation over r. Equation (7.18) can be written as

$$i\hbar\frac{1}{f(t)}\frac{\partial f(t)}{\partial t} = \frac{1}{\phi(\vec{r})}\left[-\frac{\hbar^2}{2m}\nabla^2 + \hat{V}(\vec{r})\right]\phi(\vec{r}), \qquad (7.19)$$

in which we see that both sides of the equation depend on different (independent) variables. Thus, both sides must be equal to a constant, say E:

$$i\hbar\frac{1}{f(t)}\frac{\partial f(t)}{\partial t} = E,$$

$$\frac{1}{\phi(\vec{r})}\left[-\frac{\hbar^2}{2m}\nabla^2 + \hat{V}(\vec{r})\right]\phi(\vec{r}) = E. \qquad (7.20)$$

Thus, after the separation of the variables, we get two independent ordinary differential equations

$$i\hbar\frac{\partial f(t)}{\partial t} = Ef(t), \qquad (7.21)$$

$$\left[-\frac{\hbar^2}{2m}\nabla^2 + \hat{V}(\vec{r})\right]\phi(\vec{r}) = E\phi(\vec{r}). \qquad (7.22)$$

Solution (c)

We can solve the time-dependent part, Eq. (7.21), using the method of separate variables

$$\frac{df(t)}{f(t)} = \frac{E}{i\hbar}dt. \qquad (7.23)$$

Integrating both sides over time, we get

$$\ln f(t) = -i\frac{E}{\hbar}t, \qquad (7.24)$$

which gives

$$f(t) = f(0)e^{-i\frac{E}{\hbar}t}. \qquad (7.25)$$

The time-dependent part of the wave function varies in time as an exponential function. Since the probability of finding the particle at a point $\vec{r}$ and at time t is given by the square of the absolute value of the wave function, we have

$$|f(t)|^2 = |f(0)|^2. \tag{7.26}$$

Clearly, the probability is independent of time. In other words, the probability is constant in time. In physics, quantities that do not change in time are called stationary in time. Thus, the wave function of the separable Schrödinger equation is a stationary state of the particle.

Problem 7.4

Consider the wave function

$$\Psi(x, t) = \left(Ae^{ikx} + Be^{-ikx}\right) e^{i\omega t}. \tag{7.27}$$

(a) Find the probability current corresponding to this wave function.

(b) How would you interpret the physical meaning of the parameters A and B?

Solution (a)

The probability current is defined by

$$\vec{j} = \frac{\hbar}{2im} \left(\Psi^* \nabla \Psi - \Psi \nabla \Psi^*\right). \tag{7.28}$$

Since the wave function describes a particle moving in one dimension, the x direction, the probability current for the one-dimensional case simplifies to

$$\vec{j} = \frac{\hbar}{2im} \left(\Psi^* \frac{d\Psi}{dx} - \Psi \frac{d\Psi^*}{dx}\right) \hat{i}, \tag{7.29}$$

where $\hat{i}$ is the unit vector in the x direction.

If we take the derivative

$$\frac{d\Psi(x, t)}{dx} = ik\left(Ae^{ikx} - Be^{-ikx}\right)e^{i\omega t},$$

$$\frac{d\Psi^*(x, t)}{dx} = -ik\left(A^*e^{-ikx} - B^*e^{ikx}\right)e^{-i\omega t}, \qquad (7.30)$$

we get for the probability current

$$\begin{aligned}
\vec{J} &= \frac{\hbar k}{m}\left[\left(A^*e^{-ikx} + B^*e^{ikx}\right)\left(Ae^{ikx} - Be^{-ikx}\right)\right.\\
&\quad + \left.\left(Ae^{ikx} + Be^{-ikx}\right)\left(A^*e^{-ikx} - B^*e^{ikx}\right)\right]\hat{i}\\
&= \frac{\hbar k}{m}\left(|A|^2 - |B|^2\right)\hat{i}. \qquad (7.31)
\end{aligned}$$

Solution (b)

The probability current

$$\vec{J} = \frac{\hbar k}{m}\left(|A|^2 - |B|^2\right)\hat{i} \qquad (7.32)$$

is a superposition of two currents of particles of mass m moving in opposite directions. Thus, it can be written as the sum of two currents

$$\vec{J} = \vec{J}_+ + \vec{J}_-, \qquad (7.33)$$

where

$$\vec{J}_+ = \frac{\hbar k}{m}|A|^2\,\hat{i} \qquad (7.34)$$

is a current propagating to the right, in the $+x$ direction, and

$$\vec{J}_- = -\frac{\hbar k}{m}|B|^2\,\hat{i} \qquad (7.35)$$

is a current propagating to the left, in the $-x$ direction.

Hence, A can be interpreted as the amplitude of the probability current propagating in the $+x$ direction, and B can be interpreted as the amplitude of the current propagating in the $-x$ direction.

Chapter 8

Applications of Schrödinger Equation: Potential (Quantum) Wells

Problem 8.2

Solve the stationary Schrödinger equation for a particle not bounded by any potential and show that its total energy E is not quantized.

Solution

When $\hat{V}(x) = 0$, i.e., when the particle is not bounded by any potential we can rearrange the Schrödinger equation to the form

$$\frac{d^2\phi(x)}{dx^2} = -\frac{2m}{\hbar^2}E\phi(x) = -k^2\phi(x), \qquad (8.1)$$

which is a second-order differential equation with a constant positive coefficient $k^2 = 2mE/\hbar^2$.

The solution to Eq. (8.1) is either a sine or cosine function, which in general can be written in terms of complex exponentials, such as

$$\phi(x) = Ae^{ikx} + Be^{-ikx}, \qquad (8.2)$$

where A and B are constants. Since there are no potentials that could bound the particle, the solution (8.2) is valid for all x and there are

Problems and Solutions in Quantum Physics
Zbigniew Ficek
Copyright © 2016 Pan Stanford Publishing Pte. Ltd.
ISBN 978-981-4669-36-8 (Hardcover), 978-981-4669-37-5 (eBook)
www.panstanford.com

no restrictions on k. If there are no restrictions on k, it means that there are no restrictions on $E = \hbar^2 k^2/2m$. Thus, for the particle moving in an unbounded area where the potential $\hat{V}(x) = 0$, there are no restrictions on k, which means that there are no restrictions on the energy E of the particle. Hence, E can have any value ranging from zero to $+\infty$ (continuous spectrum).

Problem 8.3

Solve the Schrödinger equation with appropriate boundary conditions for an infinite square well with the width of the well a centered at $a/2$, i.e.,

$$V(x) = 0 \quad \text{for} \quad 0 \le x \le a,$$
$$V(x) = \infty \quad \text{for} \quad x < 0 \quad \text{and} \quad x > a. \quad (8.3)$$

Check that the allowed energies are consistent with those derived in the chapter for an infinite well of width a centered at the origin. Confirm that the wave function $\phi_n(x)$ can be obtained from those found in the chapter if one uses the substitution $x \to x + a/2$.

Solution

In the regions $x < 0$ and $x > a$, the potential is infinite. Therefore, in those regions, the wave function is equal to zero. Since the wave function must be continuous at $x = 0$ and $x = a$, we have $\phi(x) = 0$ at these points.

In the region $0 \le x \le a$, the wave function is of the form

$$\phi(x) = Ae^{ikx} + Be^{-ikx}. \quad (8.4)$$

Thus, at $x = 0$, the wave function $\phi(x) = 0$ when

$$A + B = 0. \quad (8.5)$$

At $x = a$, the wave function $\phi(x) = 0$ when

$$Ae^{ika} + Be^{-ika} = 0. \quad (8.6)$$

From Eq. (8.5), we find

$$B = -A, \quad (8.7)$$

whereas from Eq. (8.6), we find

$$B = -Ae^{2ika}. \tag{8.8}$$

We have obtained two different solutions for the coefficient B. We cannot accept these two different solutions, as one of the conditions imposed on the wave function says that the wave function must be a single-value function. Therefore, we have to find a condition under which the two solutions (8.7) and (8.8) are equal. It is easy to see that the two solutions for B will be equal if

$$e^{-2ika} = 1, \tag{8.9}$$

which will be satisfied when

$$e^{-2ika} = \cos(2ka) - i\sin(2ka) = 1, \tag{8.10}$$

or when

$$\sin(2ka) = 0 \quad \text{and} \quad \cos(2ka) = 1, \tag{8.11}$$

i.e., when

$$k = n\frac{\pi}{a}, \quad \text{with} \quad n = 0, 1, 2, \ldots. \tag{8.12}$$

Since $k^2 = 2mE/\hbar^2$, we get for the energy

$$E_n = \frac{\hbar^2}{2m}k^2 = n^2\frac{\pi^2\hbar^2}{2ma^2}. \tag{8.13}$$

Comparing Eq. (8.13) with Eq. (8.21) of the textbook, we see that the expressions for the energy of the particle inside the well are the same, i.e., the energy, independent of the choice of the coordinates.

Substituting either Eq. (8.7) or (8.8) into the general solution to the wave function, Eq. (8.4), we find the wave function of the particle inside the well

$$\phi_n(x) = A\sin\left(\frac{n\pi x}{a}\right), \quad \text{with} \quad n = 1, 2, 3, \ldots, \tag{8.14}$$

where the coefficient A is found from the normalization condition

$$\int_{-\infty}^{+\infty} dx\,|\phi_n(x)|^2 = |A|^2 \int_{-\infty}^{+\infty} dx\,\sin^2\left(\frac{n\pi x}{a}\right) = 1. \tag{8.15}$$

Performing integration with the wave function $\phi_n(x)$ given by Eq. (8.14), we find $A = \sqrt{2/a}$.

Comparing the solution to the wave function, Eq. (8.14), with the solution (8.22) of the textbook, we see that the wave function (8.14) can be obtained from that of the textbook by simply substituting in Eq. (8.22), $x \rightarrow x + a/2$.

Problem 8.4

Show that as $n \to \infty$, the probability of finding a particle between x and $x + \Delta x$ inside an infinite potential well is independent of x, which is the classical expectation. This result is an example of the correspondence principle that quantum theory should give the same results as classical physics in the limit of large quantum numbers.

Solution

The probability of finding a particle between x and $x + \Delta x$ is given by

$$P = |\phi_n(x)|^2 \Delta x. \tag{8.16}$$

For a particle inside an infinite potential well, the wave function is given by Eq. (8.14), so the probability is

$$P = \frac{2}{a} \sin^2\left(\frac{n\pi x}{a}\right) \Delta x = \frac{1}{a}\left[1 - \cos\left(\frac{2n\pi x}{a}\right)\right] \Delta x. \tag{8.17}$$

When $n \to \infty$, $\cos(2n\pi x/a) \to 0$, and then

$$P \to \frac{1}{a}\Delta x. \tag{8.18}$$

Clearly, the probability is independent of x. In other words, the probability is the same for any region Δx inside the well.

Problem 8.5

As we have already learned, the exclusion of $E = 0$ as a possible value for the energy of the particle and the limitation of E to a discrete set of definite values are examples of quantum effects that have no counterpart in classical physics, where all energies, including zero, are presumed possible.

Why we do not observe these quantum effects in everyday life?

Solution

We may answer this question by looking at the expression for the energy of a particle inside a potential well given by

$$E_n = n^2 \frac{\pi^2 \hbar^2}{2ma^2},$$ (8.19)

where m is the mass of the particle and a is the width (size) of the well.

The energy difference between two neighbouring states, say n and $n-1$, is

$$E_n - E_{n-1} = (2n - 1)\frac{\pi^2 \hbar^2}{2ma^2}.$$ (8.20)

In order to distinguish the energy states, the difference between the energies of two neighbouring states should be large. It should be larger than the uncertainty of the energy of the particle. We see from the above expression that the energy difference is inversely proportional to the mass of the particle and the size of the well. These two parameters should be very small to have the energy difference large. Such small values can be achieved with small particles, such as electrons, and with structures of nano-sizes. Such objects are called microscopic objects. In everyday life, we deal with visible (macroscopic) objects, whose masses and sizes are very large compared to the mass and size of the electron. For a macroscopic object bounded in a well, the difference between the energies of the energy states is negligibly small so that a continuous rather than a discrete energy spectrum is observed.

Problem 8.6

What length scale is required to observe discrete (quantized) energies of an electron confined in an infinite potential well?

Calculate the width of the potential well in which a low-energy electron, being in the energy state $n = 2$, emits a visible light of wavelength $\lambda = 700$ nm (red) when making a transition to its ground state $n = 1$. Compare the length scale (width) to the size of an atom ~ 0.1 nm.

Solution

An electron inside an infinite potential well can have energies

$$E_n = n^2 \frac{\pi^2 \hbar^2}{2ma^2}, \qquad (8.21)$$

where m is the mass of the electron and a is the width of the well.
The energy difference between $n = 2$ and $n = 1$ is equal to

$$E_2 - E_1 = \hbar\omega, \qquad (8.22)$$

where ω is the angular frequency of light emitted. Since $\omega = 2\pi c/\lambda$, where λ is the wavelength of the emitted light, we get

$$E_2 - E_1 = \frac{3\pi^2 \hbar^2}{2ma^2} = \hbar \frac{2\pi c}{\lambda}, \qquad (8.23)$$

from which we find

$$a = \left(\frac{3\pi \hbar \lambda}{4mc} \right)^{\frac{1}{2}}. \qquad (8.24)$$

Substituting the values of the parameters

$$m = 9.11 \times 10^{-31} \text{ kg}, \quad c = 3 \times 10^8 \text{ m/s},$$
$$\hbar = 1.055 \times 10^{-34} \text{ J.s}, \quad \lambda = 700 \text{ nm} = 700 \times 10^{-9} \text{ m}, \quad (8.25)$$

we find

$$a = 0.8 \times 10^{-9} \text{ m} = 0.8 \text{ nm}. \qquad (8.26)$$

The size of the well is about eight times the size of an atom.

Problem 8.7

Particles of mass m and energy E moving in one dimension from $-x$ to $+x$ encounter a double potential step, as shown in Fig. 8.1, where

$$V_1 = \frac{\pi^2 \hbar^2}{8ma^2}, \qquad E = 2V_1, \qquad V_1 < V_2 < E. \qquad (8.27)$$

(a) Find the transmission coefficient T.
(b) Find the value of V_2 at which T is maximum.

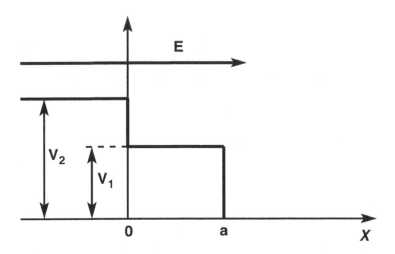

Figure 8.1 A double potential step.

Solution (a)

Since in the three regions, I: $x < 0$, II: $0 < x < a$, III: $x > a$, the energy E of the particle is larger than the potential barriers, the parameter k^2 appearing in the stationary Schrödinger equation

$$\frac{d^2\phi(x)}{dx^2} = -k^2\phi(x) \tag{8.28}$$

is a positive number, and therefore the solutions to the Schrödinger equation in these three regions are of the form

I. $\phi_1(x) = Ae^{ik_1x} + B\,e^{-ik_1x}, \qquad x < 0$

II. $\phi_2(x) = Ce^{ik_2x} + De^{-ik_2x}, \qquad 0 \le x \le a$

III. $\phi_3(x) = Fe^{ik_3x} + Ge^{-ik_3x}, \qquad x > a, \tag{8.29}$

where $k_1 = \sqrt{2m(E - V_2)}/\hbar$, $k_2 = \sqrt{2m(E - V_1)}/\hbar$, and $k_3 = \sqrt{2mE}/\hbar$.

The transmission coefficient from region I to region III is defined as

$$T = \frac{k_3}{k_1}\frac{|F|^2}{|A|^2}, \tag{8.30}$$

which can by written as

$$T = \frac{k_3}{k_1}\frac{|F|^2}{|A|^2} = \frac{k_2}{k_1}\frac{k_3}{k_2}\frac{|C|^2}{|A|^2}\frac{|F|^2}{|C|^2} = T_{12}T_{23}, \tag{8.31}$$

where

$$T_{12} = \frac{k_2}{k_1} \frac{|C|^2}{|A|^2} \qquad (8.32)$$

is the transmission coefficient from region I to region II, and

$$T_{23} = \frac{k_3}{k_2} \frac{|F|^2}{|C|^2} \qquad (8.33)$$

is the transmission coefficient from region II to region III.

We find the ratios $\frac{|C|^2}{|A|^2}$ and $\frac{|F|^2}{|C|^2}$ from the continuity conditions for the wave function and the first-order derivative at $x = 0$ and $x = a$. Since we expect that the particle transmitted to region III will move to the right (to the positive x), with no particle traveling to the left, we put $G = 0$ in the wave function in region III.

The continuity conditions for the wave function and the first-order derivative at $x = 0$ are

$$A + B = C + D, \qquad (8.34)$$
$$ik_1 A - ik_1 B = ik_2 C - ik_2 D. \qquad (8.35)$$

The continuity conditions at $x = a$ are

$$C e^{ik_2 a} + D e^{-ik_2 a} = F e^{ik_3 a}, \qquad (8.36)$$
$$ik_2 C e^{ik_2 a} - ik_2 D e^{-ik_2 a} = ik_3 F e^{ik_3 a}. \qquad (8.37)$$

The set of coupled equations (8.34) and (8.35) can be written as

$$A + B = C + D, \qquad (8.38)$$
$$A - B = \alpha(C - D), \qquad (8.39)$$

while Eqs. (8.36) and (8.37) can be written as

$$C e^{ik_2 a} + D e^{-ik_2 a} = F e^{ik_3 a}, \qquad (8.40)$$
$$C e^{ik_2 a} - D e^{-ik_2 a} = \beta F e^{ik_3 a}, \qquad (8.41)$$

where $\alpha = k_2 / k_1$ and $\beta = k_3 / k_2$.

First, we will find from Eqs. (8.40) and (8.41) the constants C and D in terms of F, which will give us the required ratio $|F|^2 / |C|^2$.

By adding Eqs. (8.40) and (8.41), we obtain

$$2C e^{ik_2 a} = (1 + \beta) F e^{ik_3 a}, \qquad (8.42)$$

from which, we find

$$C = \frac{1}{2}(1 + \beta)F e^{i(k_3 - k_2)a}.$$ (8.43)

Similarly, by subtracting Eqs. (8.40) and (8.41), we obtain

$$2D e^{-ik_2 a} = (1 - \beta)F e^{ik_3 a},$$ (8.44)

from which, we find

$$D = \frac{1}{2}(1 - \beta)F e^{i(k_3 + k_2)a}.$$ (8.45)

Thus, we find from Eq. (8.43) that

$$\frac{F}{C} = \frac{2}{(1 + \beta)} e^{-i(k_3 - k_2)a},$$ (8.46)

from which we obtain

$$\frac{|F|^2}{|C|^2} = \frac{4}{(1 + \beta)^2}.$$ (8.47)

Now, we will find the ratio $\frac{|C|^2}{|A|^2}$. By adding Eqs. (8.38) and (8.39), we obtain

$$2A = (1 + \alpha)C + (1 - \alpha)D,$$ (8.48)

and substituting for D from Eq. (8.45), we find

$$2A = (1 + \alpha)C + \frac{1}{2}(1 - \alpha)(1 - \beta)F e^{i(k_3 + k_2)a}$$

$$= (1 + \alpha)C + \frac{1}{2}(1 - \alpha)(1 - \beta)e^{i(k_3 + k_2)a}\frac{2C}{(1 + \beta)}e^{-i(k_3 - k_2)a}$$

$$= \left[(1 + \alpha) + \frac{(1 - \alpha)(1 - \beta)}{(1 + \beta)}e^{2ik_2 a}\right]C.$$ (8.49)

Thus,

$$\frac{C}{A} = \frac{2(1 + \beta)}{\left[(1 + \alpha)(1 + \beta) + (1 - \alpha)(1 - \beta)e^{2ik_2 a}\right]}.$$ (8.50)

However,

$$2k_2 a = 2\sqrt{\frac{2m}{\hbar^2}(E - V_1)}a = 2\sqrt{\frac{2m}{\hbar^2}V_1}a = 2\sqrt{\frac{2m}{\hbar^2}\frac{\pi^2\hbar^2}{8ma^2}}a = \pi.$$ (8.51)

Hence

$$e^{2ik_2 a} = e^{i\pi} = -1,$$ (8.52)

and then

$$\frac{C}{A} = \frac{2(1+\beta)}{\left[(1+\alpha)(1+\beta) + (1-\alpha)(1-\beta)e^{2ik_2a}\right]}$$

$$= \frac{2(1+\beta)}{\left[(1+\alpha)(1+\beta) - (1-\alpha)(1-\beta)\right]}$$

$$= \frac{(1+\beta)}{(\alpha+\beta)}. \tag{8.53}$$

Thus

$$\frac{|C|^2}{|A|^2} = \frac{(1+\beta)^2}{(\alpha+\beta)^2}. \tag{8.54}$$

With the solutions (8.47) and (8.54), we find the transmission coefficients

$$T_{12} = \frac{k_2}{k_1}\frac{|C|^2}{|A|^2} = \alpha\frac{(1+\beta)^2}{(\alpha+\beta)^2},$$

$$T_{23} = \frac{k_3}{k_2}\frac{|F|^2}{|C|^2} = \beta\frac{4}{(1+\beta)^2}, \tag{8.55}$$

which lead to the total transmission coefficient

$$T = T_{12}T_{23} = \frac{4\alpha\beta}{(\alpha+\beta)^2}. \tag{8.56}$$

Solution (b)

It is easy to see from the above equation that the transmission coefficient is maximum, $T = 1$, when $\alpha = \beta$, i.e., when

$$\frac{k_2}{k_1} = \frac{k_3}{k_2}, \tag{8.57}$$

from which we find that $T = 1$ when

$$k_2^2 = k_1 k_3. \tag{8.58}$$

Substituting the explicit forms of k_1, k_2, and k_3, we obtain

$$\frac{2m}{\hbar^2}(E - V_1) = \frac{2m}{\hbar^2}\sqrt{E(E - V_2)} \tag{8.59}$$

from which, we get

$$(E - V_1) = \sqrt{E(E - V_2)}. \tag{8.60}$$

Since $E = 2V_1$, we find from the above equation

$$V_1 = \sqrt{2V_1(2V_1 - V_2)} \qquad \Rightarrow \qquad V_2 = \frac{3}{2}V_1. \tag{8.61}$$

Problem 8.8

Show that the particle probability current density $\vec{J}$ is zero in region I, and deduce that $R = 1$, $T = 0$. This is the case of total reflection; the particle coming toward the barrier will eventually be found moving back. "Eventually", because the reversal of direction is not sudden. Quantum barriers are "spongy" in the sense the quantum particle may penetrate them in a way that classical particles may not.

Solution

The probability current is defined as

$$\vec{J} = \frac{\hbar}{2im} \left(\phi^* \frac{d\phi}{dx} - \phi \frac{d\phi^*}{dx} \right). \tag{8.62}$$

In region I, the wave function of the particles is

$$\phi_1(x) = C e^{k_1 x}. \tag{8.63}$$

Hence

$$\frac{d\phi_1}{dx} = k_1 C e^{k_1 x}. \tag{8.64}$$

and then

$$\phi_1^* \frac{d\phi_1}{dx} = k_1 |C|^2 e^{2k_1 x}. \tag{8.65}$$

By taking the complex conjugate of the above equation, we obtain

$$\phi_1 \frac{d\phi_1^*}{dx} = k_1 |C|^2 e^{2k_1 x}. \tag{8.66}$$

Therefore, we see that

$$\phi_1^* \frac{d\phi_1}{dx} - \phi_1 \frac{d\phi_1^*}{dx} = 0. \tag{8.67}$$

Thus

$$J_1 = \frac{\hbar}{2im} \left(\phi_1^* \frac{d\phi_1}{dx} - \phi_1 \frac{d\phi_1^*}{dx} \right) = 0, \tag{8.68}$$

i.e., in region I, the probability current is zero.

Consider now the wave function of the particle in region II

$$\phi_2(x) = A e^{ik_2 x} + B e^{-ik_2 x}. \tag{8.69}$$

The first term on the right-hand side of this equation describes particles moving to the right, away from the barrier between regions I and II, whereas the second term describes particles moving to the left, toward the barrier between I and II. Therefore, the amplitude B can be treated as an amplitude of particles incident on the barrier, while A can be treated as an amplitude of particles reflected from the barrier. Thus, we may introduce a reflection coefficient

$$R = \frac{|A|^2}{|B|^2}. \tag{8.70}$$

Using the expression for the relation between A and B, Eq. (8.97) of the textbook

$$A = \frac{(i\beta + 1)}{(i\beta - 1)} Be^{iak_2}, \tag{8.71}$$

we readily find that

$$R = \frac{|A|^2}{|B|^2} = \frac{|(i\beta + 1)|^2}{|(i\beta - 1)|^2} = \frac{(i\beta + 1)(-i\beta + 1)}{(i\beta - 1)(-i\beta - 1)}$$
$$= \frac{(i\beta + 1)(i\beta - 1)}{(i\beta - 1)(i\beta + 1)} = 1. \tag{8.72}$$

The reflection coefficient $R = 1$ even if the particles can penetrate the barrier.

Problem 8.9

Recall the case of $E > V_0$, discussed briefly in Section 8.3.1 of the textbook.

(a) Evaluate the transmission coefficient from region I to region III.
(b) Under which condition, the transmission coefficient becomes unity ($T = 1$)?

Solution (a)

The general solution to the Schrödinger equation for the wave function of the particle with energy $E > V_0$ is of the form

$$\text{I.} \quad \phi_1(x) = C e^{ik_1 x} + D e^{-ik_1 x}, \qquad x < -\frac{a}{2}$$

$$\text{II.} \quad \phi_2(x) = A e^{ik_2 x} + B e^{-ik_2 x}, \qquad -\frac{a}{2} \le x \le \frac{a}{2}$$

$$\text{III.} \quad \phi_3(x) = F e^{ik_1 x}, \qquad x > \frac{a}{2}. \tag{8.73}$$

The transmission coefficient from region I to region III is given by the ratio $T = |F|^2/|C|^2$. Thus, we will try to find the coefficient F in terms of C using the properties of the wave function: $\phi(x)$ and the first-order derivative $d\phi(x)/dx$ must be finite and continuous everywhere, in particular, at the boundaries $x = -a/2$ and $x = a/2$.

The first-order derivatives of the wave function in different regions are

$$\text{I.} \quad \frac{d\phi_1}{dx} = ik_1 C e^{ik_1 x} - ik_1 D e^{-ik_1 x},$$

$$\text{II.} \quad \frac{d\phi_2}{dx} = ik_2 A e^{ik_2 x} - ik_2 B e^{-ik_2 x},$$

$$\text{III.} \quad \frac{d\phi_3}{dx} = ik_1 F e^{ik_1 x}. \tag{8.74}$$

From the continuity of the wave function and the first derivatives at $x = -a/2$, we get

$$C e^{-\frac{1}{2}iak_1} + D e^{\frac{1}{2}iak_1} = A e^{-i\frac{1}{2}ak_2} + B e^{i\frac{1}{2}ak_2},$$

$$ik_1 C e^{-\frac{1}{2}iak_1} - ik_1 D e^{\frac{1}{2}iak_1} = ik_2 A e^{-i\frac{1}{2}ak_2} - ik_2 B e^{i\frac{1}{2}ak_2}, \tag{8.75}$$

which can be written as

$$C e^{-\frac{1}{2}iak_1} + D e^{\frac{1}{2}iak_1} = A e^{-i\frac{1}{2}ak_2} + B e^{i\frac{1}{2}ak_2},$$

$$C e^{-\frac{1}{2}iak_1} - D e^{\frac{1}{2}iak_1} = \beta A e^{-i\frac{1}{2}ak_2} - \beta B e^{i\frac{1}{2}ak_2}, \tag{8.76}$$

where $\beta = k_2/k_1$.

By adding the two equations in (8.76), we get C in terms of A and B:

$$2C e^{-\frac{1}{2}iak_1} = (\beta + 1) A e^{-i\frac{1}{2}ak_2} + (1 - \beta) B e^{i\frac{1}{2}ak_2}, \tag{8.77}$$

From the continuity of the wave function and the first derivatives at $x = a/2$, we get

$$Fe^{\frac{1}{2}iak_1} = Ae^{i\frac{1}{2}ak_2} + Be^{-i\frac{1}{2}ak_2},$$
$$ik_1 Fe^{\frac{1}{2}iak_1} = ik_2 Ae^{i\frac{1}{2}ak_2} - ik_2 Be^{-i\frac{1}{2}ak_2}, \tag{8.78}$$

which can be written as

$$Fe^{\frac{1}{2}iak_1} = Ae^{i\frac{1}{2}ak_2} + Be^{-i\frac{1}{2}ak_2},$$
$$Fe^{\frac{1}{2}iak_1} = \beta Ae^{i\frac{1}{2}ak_2} - \beta Be^{-i\frac{1}{2}ak_2}. \tag{8.79}$$

Since the left-hand sides of the above equations are equal, we get

$$Ae^{i\frac{1}{2}ak_2} + Be^{-i\frac{1}{2}ak_2} = \beta Ae^{i\frac{1}{2}ak_2} - \beta Be^{-i\frac{1}{2}ak_2}, \tag{8.80}$$

from which we find B in terms of A as

$$B = \frac{(\beta - 1)}{(\beta + 1)} Ae^{iak_2}. \tag{8.81}$$

Substituting Eq. (8.81) in either of the expressions in (8.79), we get F in terms of A:

$$F = Aue^{i\frac{1}{2}a(k_2 - k_1)}, \tag{8.82}$$

where $u = 2\beta/(\beta + 1)$.

Substituting Eq. (8.81) into Eq. (8.77), we get C in terms of A:

$$C = \frac{1}{2} Ae^{\frac{1}{2}ia(k_1 + k_2)} \left[(\beta + 1)e^{-iak_2} - \frac{(1 - \beta)^2}{(\beta + 1)} e^{iak_2} \right]. \tag{8.83}$$

For the transmission coefficient, we need $|F|^2$ and $|C|^2$. From Eq. (8.82), we have

$$|F|^2 = |A|^2 u^2, \tag{8.84}$$

and from Eq. (8.83), we have

$$|C|^2 = \frac{1}{4}|A|^2 \left\{ 4u^2 + 2(\beta - 1)^2 [1 - \cos(2ak_2)] \right\}. \tag{8.85}$$

Thus, the transmission coefficient is of the form

$$T = \frac{|F|^2}{|C|^2} = \frac{4u^2}{4u^2 + 2(\beta - 1)^2 [1 - \cos(2ak_2)]}. \tag{8.86}$$

Note that in general $T < 1$, showing that even when $E > V_0$, not all particles can be transmitted to region III, a part of the particles can be reflected back to region I.

Solution (b)

Lets look more closely at the expression for the transmission coefficient derived in part (a):

$$T = \frac{4u^2}{4u^2 + 2(\beta - 1)^2 \left[1 - \cos(2ak_2)\right]}. \tag{8.87}$$

It is easily seen that there are two different conditions for $T = 1$. The first one is a trivial, $\beta = 1$ or equivalently $k_1 = k_2$, and corresponds to the situation of $V_0 = 0$, i.e., there is no potential barrier. The second condition is more interesting and corresponds to

$$\cos(2ak_2) = 1, \tag{8.88}$$

which happens when

$$2ak_2 = 2\pi n, \quad n = 0, 1, 2, \ldots \tag{8.89}$$

or when the energy of the particle satisfies the conditions

$$E = n^2 \frac{\pi^2 \hbar^2}{2ma^2}. \tag{8.90}$$

Thus, for some discrete energies $E > V_0$, the transmission coefficient from region I to region III equals $T = 1$, independent of the value of V_0.

Problem 8.10

A rectangular potential well is bounded by a wall of infinite high on one side and a wall of high V_0 on the other, as shown in Figure 8.2. The well has a width a, and a particle located inside the well has energy $E < V_0$.

(a) Find the wave function of the particle inside the well.
(b) Show that the energy of the particle is quantized.
(c) Discuss the dependence of the number of energy levels inside the well on V_0.

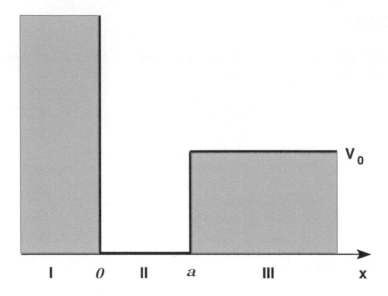

Figure 8.2 Potential well of semi-infinite depth.

Solution (a)

The general solution to the Schrödinger equation for the wave function of a particle located in region II and with energy $E < V_0$ is of the form

$$\text{I.} \quad \phi_1(x) = 0, \qquad x < 0$$
$$\text{II.} \quad \phi_2(x) = Ae^{ik_2x} + Be^{-ik_2x}, \qquad 0 \le x \le a$$
$$\text{III.} \quad \phi_3(x) = Fe^{-k_3x}, \qquad x > a, \qquad (8.91)$$

where $k_2 = \sqrt{2mE}/\hbar$, and $k_3 = \sqrt{2m(V_0 - E)}/\hbar$.

At $x = 0$, the wave function is continuous, $\phi_1(0) = \phi_2(0)$, when

$$A + B = 0. \qquad (8.92)$$

Hence

$$B = -A. \qquad (8.93)$$

Thus, the wave function of the particle inside the well is of the form

$$\phi_2(x) = A\left(e^{ik_2x} - e^{-ik_2x}\right) = 2iA\sin(k_2a), \qquad (8.94)$$

where the coefficient A is found from the normalization condition.

Solution (b)

Consider the continuity conditions for the wave function and the first-order derivatives at $x = a$:

$$\phi_2(a) = \phi_3(a),$$

$$\left.\frac{d\phi_2(x)}{dx}\right|_{x=a} = \left.\frac{d\phi_3(x)}{dx}\right|_{x=a}. \tag{8.95}$$

The above continuity conditions lead to two equations for F:

$$Fe^{-k_3 a} = Ae^{ik_2 a} + Be^{-ik_2 a},$$

$$-k_3 Fe^{-k_3 a} = ik_2 Ae^{ik_2 a} - ik_2 Be^{-ik_2 a}, \tag{8.96}$$

which, after substituting $B = -A$, take the form

$$Fe^{-k_3 a} = Ae^{ik_2 a} - Ae^{-ik_2 a},$$

$$Fe^{-k_3 a} = -i\beta \left(Ae^{ik_2 a} + Ae^{-ik_2 a}\right), \tag{8.97}$$

where $\beta = k_2/k_3$.

Since $e^{ik_2 a} - e^{-ik_2 a} = 2i \sin(k_2 a)$ and $e^{ik_2 a} + e^{-ik_2 a} = 2 \cos(k_2 a)$, the above equations can be simplified to

$$Fe^{-k_3 a} = 2i A \sin(k_2 a),$$

$$Fe^{-k_3 a} = -2i\beta A \cos(k_2 a). \tag{8.98}$$

Thus, we have two different solutions to F. However, we cannot accept both the solutions as it would mean that there are two different probabilities of finding the particle at a point x inside region III. Therefore, we have to find under which circumstances these two solutions are equal.

It is easily seen from Eq. (8.98) that the two solutions to F will be equal when

$$\sin(k_2 a) = -\beta \cos(k_2 a), \tag{8.99}$$

which can be written as

$$\tan(k_2 a) = -\beta = -\frac{k_2}{k_3}. \tag{8.100}$$

Introduce a notation

$$ak_2 = \sqrt{\frac{2ma^2 E}{\hbar^2}} = \varepsilon,$$

$$ak_3 = \sqrt{\frac{2ma^2(V_0 - E)}{\hbar^2}} = \sqrt{\frac{2ma^2 V_0}{\hbar^2} - \frac{2ma^2 E}{\hbar^2}} = \sqrt{\eta^2 - \varepsilon^2}, \tag{8.101}$$

where $\eta = \sqrt{2ma^2 V_0/\hbar^2}$. Hence, Eq. (8.100) can be written as

$$\tan \varepsilon = -\frac{\varepsilon}{\sqrt{\eta^2 - \varepsilon^2}} \qquad (8.102)$$

or the a form

$$\sqrt{\eta^2 - \varepsilon^2} = -\varepsilon \cot \varepsilon. \qquad (8.103)$$

Equation (8.103) is transcendental, so the exact solution can only be found numerically.

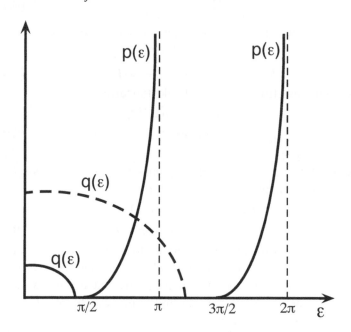

Figure 8.3 Graphical solution to Eq. (8.103) versus ε for two different values of η: $\eta < \pi/2$ (solid line), $\eta < 3\pi/2$ (dashed line).

To obtain the solution graphically, we plot in Fig. 8.3 the functions

$$p(\varepsilon) = -\varepsilon \cot \varepsilon, \quad q(\varepsilon) = \sqrt{\eta^2 - \varepsilon^2}, \qquad (8.104)$$

against ε for two different values of η. Whenever the $q(\varepsilon)$ curve crosses the $-\cot \varepsilon$ curve in Fig. 8.3, we have a solution which satisfies the necessary condition that F is a single-value amplitude. We see from the figure that the equation $p(\varepsilon) = q(\varepsilon)$ is satisfied only for discrete values of ε. Since the energy E is proportional to ε, as seen from Eq. (8.101), we find that the energy of the particle is quantized in region II.

Solution (c)

It is easily seen from Fig. 8.3 that the number of crossing points of $q(\varepsilon)$ with $p(\varepsilon)$ depends on the value of η. Note that the number of crossings corresponds to a number of energy states fitted into the well, region II. Thus, there is no crossing point when $\eta < \pi/2$. When $\eta < 3\pi/2$ there is one crossing point, for $\eta < 5\pi/2$ there are two crossing points, and so on.

It is interesting to note that there is a possibility of no energy state allowed inside the well. Since, $\eta = \sqrt{2ma^2 V_0/\hbar^2}$, we see that for

$$V_0 < \frac{\pi^2 \hbar^2}{8ma^2},$$ (8.105)

there is no energy state allowed inside the well.

Problem 8.15

Particles of mass m and energy $E < V_0$ moving in one dimension from $-x$ to $+x$ encounter a non-symmetric barrier, as shown in Fig. 8.4.

(a) Find the transmission coefficient T.
(b) Show that in the limit of $a \rightarrow 0$, the transmission coefficient reduces to that of the step potential.
(c) Does the transmission coefficient depend on the direction of propagation of the particles?

Solution (a)

Since in regions I: $x < -a$ and III: $x > a$, the energy E of the particle is larger than the potential barriers, the parameter k^2 appearing in the stationary Schrödinger equation is a positive number, and therefore the solutions to the Schrödinger equation in regions I and

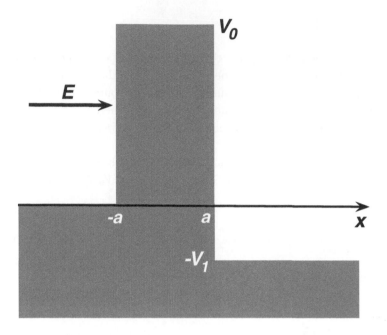

Figure 8.4   Tunneling through a non-symmetric barrier.

III are of the form

$$\text{I.} \quad \phi_1(x) = Ae^{ik_1x} + Be^{-ik_1x}, \qquad x < -a$$
$$\text{III.} \quad \phi_3(x) = Fe^{ik_3x} + Ge^{-ik_3x}, \qquad x > a, \qquad (8.106)$$

where $k_1 = \sqrt{2mE}/\hbar$ and $k_3 = \sqrt{2m(E+V_1)}/\hbar$. The transmitted particle to the region III will move to the right (to the positive x), so we do not expect any particles travelling to the left. Thus, we put $G = 0$ in the wave function in the region III.

In region II: $-a \leq x \leq a$, the energy E of the particle is smaller than the potential barrier. Therefore, the solution to the Schrödinger equation in region II is of the form

$$\text{II.} \quad \phi_2(x) = Ce^{-k_2x} + De^{k_2x}, \qquad (8.107)$$

where $k_2 = \sqrt{2m(V_0 - E)}/\hbar$.

The transmission coefficient from region I to region III is defined as

$$T = \frac{k_3}{k_1}\frac{|F|^2}{|A|^2}, \qquad (8.108)$$

where the ratio $|F|^2/|A|^2$ is found from the continuity conditions for the wave function and its first-order derivatives at $x = -a$ and $x = a$.

The continuity conditions at $x = -a$ are

$$Ae^{-ik_1a} + Be^{ik_2a} = Ce^{k_2a} + De^{-k_2a},$$
$$ik_1 Ae^{-ik_1a} - ik_1 Be^{ik_1a} = -k_2 Ce^{k_2a} + k_2 De^{-k_2a}. \quad (8.109)$$

The above set of coupled equations, can be written as

$$Ae^{-ik_1a} + Be^{ik_2a} = Ce^{k_2a} + De^{-k_2a}, \quad (8.110)$$
$$Ae^{-ik_1a} - Be^{ik_1a} = i\beta Ce^{k_2a} - i\beta De^{-k_2a}. \quad (8.111)$$

where $\beta = k_2/k_1$.

The continuity conditions at $x = a$ are

$$Ce^{-k_2a} + De^{k_2a} = Fe^{ik_3a},$$
$$-k_2 Ce^{-k_2a} + k_2 De^{k_2a} = ik_3 Fe^{ik_3a}, \quad (8.112)$$

which we can write as

$$Ce^{-k_2a} + De^{k_2a} = Fe^{ik_3a}, \quad (8.113)$$
$$-Ce^{-k_2a} + De^{k_2a} = i\gamma Fe^{ik_3a}, \quad (8.114)$$

where $\gamma = k_3/k_2$.

By adding Eqs. (8.110) and (8.111), we obtain

$$2Ae^{-ik_1a} = uCe^{k_2a} + u^* De^{-k_2a}, \quad (8.115)$$

where $u = 1 + i\beta$ and $u^* = 1 - i\beta$.

By adding and subtracting Eqs. (8.113) and (8.114), we find the coefficients D and C in terms of F:

$$2De^{k_2a} = wFe^{ik_3a}, \quad (8.116)$$
$$2Ce^{-k_2a} = w^* Fe^{ik_3a}, \quad (8.117)$$

where $w = 1 + i\gamma$ and $w^* = 1 - i\gamma$.

Substituting Eqs. (8.116) and (8.117) into Eq. (8.115), we obtain A in terms of F:

$$4Ae^{-ik_1a} = ue^{2k_2a}w^* Fe^{ik_3a} + u^*e^{-2k_2a}wFe^{ik_3a}, \quad (8.118)$$

from which we find

$$4Ae^{-i(k_1+k_3)a} = F\left[uw^*e^{2k_2a} + u^*we^{-2k_2a}\right]. \quad (8.119)$$

Thus

$$\frac{|F|^2}{|A|^2} = \frac{16}{|uw^* e^{2k_2 a} + u^* we^{-2k_2 a}|^2}. \tag{8.120}$$

Let us simplify the denominator on the right-hand side of the above equation. It can be written as

$$|uw^* \, e^{2k_2 a} + u^* we^{-2k_2 a}|^2$$

$$= (uw^* e^{2k_2 a} + u^* we^{-2k_2 a})(u^* we^{2k_2 a} + uw^* e^{-2k_2 a})$$

$$= |u|^2 |w|^2 e^{4k_2 a} + (u^2 w^{*2} + u^{*2} w^2) + |u|^2 |w|^2 e^{-4k_2 a}$$

$$= |u|^2 |w|^2 (e^{4k_2 a} + e^{-4k_2 a}) + (u^2 w^{*2} + u^{*2} w^2)$$

$$= 2|u|^2 |w|^2 \cosh(4k_2 a) + (uw^*)^2 + (u^* w)^2. \tag{8.121}$$

Since $\cosh(4k_2 a) = 1 + 2\sinh^2(2k_2 a)$, we obtain

$$\frac{|F|^2}{|A|^2} = \frac{16}{(uw^* + u^* w)^2 + 4|u|^2 |w|^2 \sinh^2(2k_2 a)}, \tag{8.122}$$

and then the transmission coefficient can be written as

$$T = \frac{k_3}{k_1} \frac{|F|^2}{|A|^2} = \frac{P}{1 + Q\sinh^2(2k_2 a)}, \tag{8.123}$$

where

$$P = \frac{k_3}{k_1} \frac{16}{(uw^* + u^* w)^2}, \tag{8.124}$$

$$Q = \frac{4|u|^2 |w|^2}{(uw^* + u^* w)^2}. \tag{8.125}$$

In terms of the constants k_1, k_2, and k_3, the coefficients P and Q are

$$P = \frac{k_3}{k_1} \frac{16}{(uw^* + u^* w)^2} = \frac{k_3}{k_1} \frac{16}{4(1 + \beta\gamma)^2} = \frac{4k_3 k_1}{(k_1 + k_3)^2}. \tag{8.126}$$

and

$$Q = \frac{4|u|^2 |w|^2}{(uw^* + u^* w)^2} = \frac{4(1 + \beta^2)(1 + \gamma^2)}{4(1 + \beta\gamma)^2} = \frac{(k_1 + k_2)^2 (k_2 + k_3)^2}{k_2^2 (k_1 + k_3)^2}. \tag{8.127}$$

Solution (b)

In the limit of $a \to 0$, the function $\sinh^2(2k_2 a) \to 0$, and then the transmission coefficient reduces to $T = P$, i.e.,

$$T = \frac{4k_3 k_1}{(k_1 + k_3)^2}. \tag{8.128}$$

This is precisely the result obtained for the step potential. For details see Section 8.2 of the textbook Eq. (8.75), page 126.

Solution (c)

No, in terms of the constants k, the change in the direction of propagation is equivalent to exchange $k_3 \leftrightarrow k_1$. We see from the above that in this case, the coefficients P and Q remain the same.

This is an important result. Even though the barrier and energy structure do not appear symmetrical, the barrier is a linear, passive structure. Therefore, the transmission should be the same regardless of the direction from which one approaches the barrier.

Chapter 9

Multidimensional Quantum Wells

Problem 9.1

This problem illustrates why the tunneling (flow) of an electron between different quantum dots is possible only for specific (discrete) energies of the electron.

Consider a simplified situation, a one-dimensional system of quantum wells, as shown in Fig. 9.1. The well represents a quantum dot. Show, using the method we learned in the previous chapter on applications of the Schrödinger equation, that an electron of energy $E < V_0$ and being in region I can tunnel through the quantum well (region II) to region III only if E is equal to the energy of one of the discrete energy levels inside the well.

Solution

In fact, we can distinguish here five different regions for the wave functions. The general solution to the Schrödinger equation for the wave function of a particle in region I traveling to the right and with

Problems and Solutions in Quantum Physics
Zbigniew Ficek
Copyright © 2016 Pan Stanford Publishing Pte. Ltd.
ISBN 978-981-4669-36-8 (Hardcover), 978-981-4669-37-5 (eBook)
www.panstanford.com

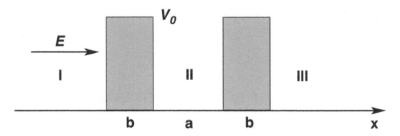

Figure 9.1 Tunneling through a quantum well of thickness a formed by two barriers, each of thickness b and of finite potential V_0.

energy $E < V_0$ is of the form

$$
\begin{aligned}
\text{I.} \quad & \phi_1(x) = Ae^{ik_1x} + Be^{-ik_1x}, && x < 0 \\
\text{B1.} \quad & \phi_2(x) = Ce^{k_2x} + De^{-k_2x}, && 0 \le x \le b \\
\text{II.} \quad & \phi_3(x) = Fe^{ik_1x} + Ge^{-ik_1x}, && b \le x \le b + a \\
\text{B2.} \quad & \phi_4(x) = He^{k_2x} + Ue^{-k_2x}, && b + a \le x \le 2b + a \\
\text{III.} \quad & \phi_5(x) = We^{ik_1x}, && x > 2b + a, \quad (9.1)
\end{aligned}
$$

where $k_1 = \sqrt{2mE}/\hbar$ and $k_2 = \sqrt{2m(V_0 - E)}/\hbar$.

Let us first find the transmission coefficient from region I to region III. It is given by $T = |W|^2/|A|^2$.

The relation between W and A is found using the boundary conditions that the wave function and its first-order derivatives must be continuous at the boundaries $x = 0, b, b + a, 2b + a$. The continuity conditions at $x = 0$ are

$$
\begin{aligned}
A + B &= C + D, \\
A - B &= -i\beta \left(C - D \right), \quad (9.2)
\end{aligned}
$$

where $\beta = k_2/k_1$. Hence, by eliminating B between these equations, we get A in terms of C and D:

$$
A = \frac{1}{2}(1 - i\beta)C + \frac{1}{2}(1 + i\beta)D. \quad (9.3)
$$

The continuity conditions at $x = b$ are

$$
\begin{aligned}
Ce^{k_2b} + De^{-k_2b} &= Fe^{ik_1b} + Ge^{-ik_1b}, \\
Ce^{k_2b} - De^{-k_2b} &= \frac{i}{\beta}\left(Fe^{ik_1b} - Ge^{-ik_1b} \right), \quad (9.4)
\end{aligned}
$$

from which we find C and D in terms of F and G

$$2C e^{k_2 b} = \left(1 + \frac{i}{\beta}\right) F e^{ik_1 b} + \left(1 - \frac{i}{\beta}\right) G e^{-ik_1 b},$$

$$2D e^{-k_2 b} = \left(1 - \frac{i}{\beta}\right) F e^{ik_1 b} + \left(1 + \frac{i}{\beta}\right) G e^{-ik_1 b}. \qquad (9.5)$$

Hence

$$C = \frac{i}{2\beta} \left[(1 - i\beta) F e^{ik_1 b} - (1 + i\beta) G e^{-ik_1 b}\right] e^{-k_2 b},$$

$$D = -\frac{i}{2\beta} \left[(1 + i\beta) F e^{ik_1 b} - (1 - i\beta) G e^{-ik_1 b}\right] e^{k_2 b}. \qquad (9.6)$$

Substituting into Eq. (9.3), we get A in terms of F and G:

$$A = -\frac{i}{2\beta} \left\{ \left[(1 - \beta^2) \sinh(k_2 b) + 2i\beta \cosh(k_2 b)\right] F e^{ik_1 b} \right.$$

$$\left. - (1 + \beta^2) \sinh(k_2 b) G e^{-ik_1 b} \right\}, \qquad (9.7)$$

where we have used the relations

$$\cosh\alpha = \frac{1}{2}\left(e^\alpha + e^{-\alpha}\right), \quad \sinh\alpha = \frac{1}{2}\left(e^\alpha - e^{-\alpha}\right), \quad \alpha = k_2 b.$$

$$(9.8)$$

The continuity conditions at $x = b + a$ are

$$F e^{ik_1(b+a)} + G e^{-ik_1(b+a)} = H e^{k_2(b+a)} + U e^{-k_2(b+a)},$$

$$F e^{ik_1(b+a)} - G e^{-ik_1(b+a)} = -i\beta \left(H e^{k_2(b+a)} - U e^{-k_2(b+a)}\right), \qquad (9.9)$$

which give

$$F e^{ik_1 b} = \frac{1}{2} \left[(1 - i\beta) H e^{k_2(b+a)} + (1 + i\beta) U e^{-k_2(b+a)}\right] e^{-ik_1 a},$$

$$G e^{-ik_1 b} = \frac{1}{2} \left[(1 + i\beta) H e^{k_2(b+a)} + (1 - i\beta) U e^{-k_2(b+a)}\right] e^{ik_1 a}. \qquad (9.10)$$

The continuity conditions at $x = 2b + a$ are

$$H e^{k_2(2b+a)} + U e^{-k_2(2b+a)} = W e^{ik_1(2b+a)},$$

$$H e^{k_2(2b+a)} - U e^{-k_2(2b+a)} = \frac{i}{\beta} W e^{ik_1(2b+a)}, \qquad (9.11)$$

from which we find H and U in terms of W:

$$H e^{k_2(b+a)} = \frac{i}{2\beta} (1 - i\beta) W e^{ik_1(2b+a) - k_2 b},$$

$$U e^{-k_2(b+a)} = -\frac{i}{2\beta} (1 + i\beta) W e^{ik_1(2b+a) + k_2 b}. \qquad (9.12)$$

Substituting Eq. (9.12) into Eq. (9.10), we get

$$F e^{ik_1 b} = -\frac{i}{2\beta} \left[(1 - \beta^2) \sinh(k_2 b) + 2i\beta \cosh(k_2 b) \right] W e^{2ik_1 b},$$

$$G e^{-ik_1 b} = -\frac{i}{2\beta} (1 + \beta^2) \sinh(k_2 b) W e^{2ik_1(a+b)}. \qquad (9.13)$$

Substituting Eq. (9.13) into Eq. (9.7), we get A in terms of W:

$$A = -\frac{1}{4\beta^2} \left[(1 - \beta^2) \sinh(k_2 b) + 2i\beta \cosh(k_2 b) \right]^2 W e^{2ik_1 b}$$

$$+ \frac{1}{4\beta^2} (1 + \beta^2)^2 \sinh^2(k_2 b) W e^{2ik_1(a+b)}, \qquad (9.14)$$

which can be written as

$$4\beta^2 \frac{A e^{-2ik_1 b}}{W} = - \left[(1 - \beta^2) \sinh(k_2 b) + 2i\beta \cosh(k_2 b) \right]^2$$

$$+ (1 + \beta^2)^2 \sinh^2(k_2 b) e^{2ik_1 a}$$

$$= 4\beta^2 \left[\cosh(k_2 b) + i \frac{(\beta^2 - 1)}{2\beta} \sinh(k_2 b) \right]^2$$

$$+ 4\beta^2 \left(\frac{1 + \beta^2}{2\beta} \right)^2 \sinh^2(k_2 b) e^{2ik_1 a}. \qquad (9.15)$$

Hence, the ratio W/A required for the transmission coefficient is of the form

$$\frac{W}{A} = \frac{e^{-2ik_1 b}}{(\cosh \alpha + i\gamma \sinh \alpha)^2 + \eta^2 e^{2iu} \sinh^2 \alpha}, \qquad (9.16)$$

where for clarity of expression, we have introduced the notations

$$\alpha = k_2 b, \quad u = k_1 a, \quad \gamma = \frac{\beta^2 - 1}{2\beta}, \quad \eta = \frac{\beta^2 + 1}{2\beta}. \qquad (9.17)$$

Since the denominator in Eq. (9.16) is the sum of squares of two terms, we may use the relation $a^2 + b^2 = (a + ib)(a - ib)$ and write the ratio W/A as

$$\frac{W}{A} = \frac{e^{-2ik_1 b}}{\left[\cosh \alpha + i(\gamma + \eta e^{iu}) \sinh \alpha \right] \left[\cosh \alpha + i(\gamma - \eta e^{iu}) \sinh \alpha \right]}. \qquad (9.18)$$

It is not difficult to show that the ratio is maximum when $e^{iu} = \pm 1$. This happens when

$$u = k_1 a = n\pi, \quad n = 0, 1, 2, \ldots \qquad (9.19)$$

Since $k_1 = \sqrt{2mE}/\hbar$, we have evidence that the condition (9.19) corresponds to discrete energy levels of the well:

$$k_1 a = n\pi \Rightarrow \sqrt{\frac{2mE}{\hbar^2}} a = n\pi \Rightarrow E = n^2 \frac{\pi^2 \hbar^2}{2ma^2}. \qquad (9.20)$$

Thus, under the condition that the energy of the electron corresponds to the energy levels of the well, the transmission through the well is maximum with the transmission rate

$$|T|^2 = \frac{1}{\left[\cosh^2 \alpha + (\gamma + \eta)^2 \sinh^2 \alpha\right]\left[\cosh^2 \alpha + (\gamma - \eta)^2 \sinh^2 \alpha\right]}$$

$$= \frac{1}{1 + \eta^2 \sinh^2(2\alpha)}, \qquad (9.21)$$

where we have used the relation $\cosh^2 \alpha = 1 + \sinh^2 \alpha$.

Problem 9.2

Find the number of wave functions (energy states) of a particle in a quantum well of sides of equal lengths corresponding to energy

$$E = \frac{9\pi^2 \hbar^2}{2ma^2}, \qquad (9.22)$$

i.e., for the combination of n_1, n_2, and n_3 whose squares sum to 9.

Solution

To obtain the number of wave functions, we need to find the number of different trios of integers (n_1, n_2, n_3) whose squares sum to 9:

$$n_1^2 + n_2^2 + n_3^2 = 9. \qquad (9.23)$$

This can be obtained from

n_1	n_2	n_3	
2	2	1	
2	1	2	
1	2	2	(9.24)

Thus, there are three wave functions corresponding to the energy (9.22).

Problem 9.3

Find all energy states of a particle confined inside a three-dimensional box with energies

$$E = 15\frac{\pi^2\hbar^2}{2ma^2}.$$ (9.25)

Indicate the degeneracy of each energy level.

Solution

The energy states are characterized by a trio of integers n_1, n_2, n_3 whose sum of squares does not exceed 15.

$$E = (n_1^2 + n_2^2 + n_3^2)E_0 \leq 15E_0,$$ (9.26)

where $E_0 = \pi^2\hbar^2/(2ma^2)$.

The number of energy states, their energies, and their degeneracy are obtained with the following combinations of the integer numbers. Note that n_i cannot exceed 3 to have the sum of squares not exceeding 15. The energy levels of the corresponding energies are

E/E_0	n_1	n_2	n_3	Degeneracy	
3	1	1	1	1	
6	2	1	1	3	
9	2	2	1	3	
11	3	1	1	3	
12	2	2	2	1	
14	3	2	1	6	(9.27)

Thus, there are six energy states whose energies do not exceed $E = 15E_0$. Each energy state is characterized by a trio of integers, which may be rearranged to give another state of the same energy. For example, the trio (1, 1, 1) cannot be rearranged, so the lowest energy state of energy $E = 3E_0$ is a singlet. The trio (2, 1, 1) can be rearranged to (1, 2, 1) and (1, 1, 2), so the state of energy $E = 6E_0$ is a triplet.

Chapter 10

Linear Operators and Their Algebra

Problem 10.2

Let $\hat{A}$, $\hat{B}$, $\hat{C}$ be arbitrary linear operators. Prove that

(a) $\left[\hat{A}\hat{B}, \hat{C}\right] = \left[\hat{A}, \hat{C}\right]\hat{B} + \hat{A}\left[\hat{B}, \hat{C}\right]$,

(b) $\left[\hat{A}, \left[\hat{B}, \hat{C}\right]\right] + \left[\hat{B}, \left[\hat{C}, \hat{A}\right]\right] + \left[\hat{C}, \left[\hat{A}, \hat{B}\right]\right] = 0$.

Solution (a)

The left-hand side of the relation can be written as

$$\left[\hat{A}\hat{B}, \hat{C}\right] = \hat{A}\hat{B}\hat{C} - \hat{C}\hat{A}\hat{B}. \tag{10.1}$$

Consider now the right-hand side of the relation

$$\begin{aligned}
\left[\hat{A}, \hat{C}\right]\hat{B} + \hat{A}\left[\hat{B}, \hat{C}\right] &= (\hat{A}\hat{C} - \hat{C}\hat{A})\hat{B} + \hat{A}(\hat{B}\hat{C} - \hat{C}\hat{B}) \\
&= \hat{A}\hat{C}\hat{B} - \hat{C}\hat{A}\hat{B} + \hat{A}\hat{B}\hat{C} - \hat{A}\hat{C}\hat{B} \\
&= \hat{A}\hat{B}\hat{C} - \hat{C}\hat{A}\hat{B} = L, \tag{10.2}
\end{aligned}$$

as required.

Problems and Solutions in Quantum Physics
Zbigniew Ficek
Copyright © 2016 Pan Stanford Publishing Pte. Ltd.
ISBN 978-981-4669-36-8 (Hardcover), 978-981-4669-37-5 (eBook)
www.panstanford.com

Solution (b)

Consider each of the commutators separately

(i) $[\hat{A}, [\hat{B}, \hat{C}]] = [\hat{A}, (\hat{B}\hat{C} - \hat{C}\hat{B})] = \hat{A}(\hat{B}\hat{C} - \hat{C}\hat{B}) - (\hat{B}\hat{C} - \hat{C}\hat{B})\hat{A}$
$$= \hat{A}\hat{B}\hat{C} - \hat{A}\hat{C}\hat{B} - \hat{B}\hat{C}\hat{A} + \hat{C}\hat{B}\hat{A}.$$

$$(10.3)$$

(ii) $[\hat{B}, [\hat{C}, \hat{A}]] = [\hat{B}, (\hat{C}\hat{A} - \hat{A}\hat{C})] = \hat{B}(\hat{C}\hat{A} - \hat{A}\hat{C}) - (\hat{C}\hat{A} - \hat{A}\hat{C})\hat{B}$
$$= \hat{B}\hat{C}\hat{A} - \hat{B}\hat{A}\hat{C} - \hat{C}\hat{A}\hat{B} + \hat{A}\hat{C}\hat{B}.$$

$$(10.4)$$

(iii) $[\hat{C}, [\hat{A}, \hat{B}]] = [\hat{C}, (\hat{A}\hat{B} - \hat{B}\hat{A})] = \hat{C}(\hat{A}\hat{B} - \hat{B}\hat{A}) - (\hat{A}\hat{B} - \hat{B}\hat{A})\hat{C}$
$$= \hat{C}\hat{A}\hat{B} - \hat{C}\hat{B}\hat{A} - \hat{A}\hat{B}\hat{C} + \hat{B}\hat{A}\hat{C}.$$

$$(10.5)$$

It is easy to see that the sum (i) + (ii) + (iii) = 0, as required.

Problem 10.3

Let

$$[\hat{A}, \hat{B}] = i\hbar\hat{C} \quad \text{and} \quad [\hat{B}, \hat{C}] = i\hbar\hat{A}. \qquad (10.6)$$

Show that

$$\hat{B}\left(\hat{C} + i\hat{A}\right) = \left(\hat{C} + i\hat{A}\right)\left(\hat{B} + \hbar\right),$$
$$\hat{B}\left(\hat{C} - i\hat{A}\right) = \left(\hat{C} - i\hat{A}\right)\left(\hat{B} - \hbar\right). \qquad (10.7)$$

Solution

Since $[\hat{A}, \hat{B}] = i\hbar\hat{C}$ and $[\hat{B}, \hat{C}] = i\hbar\hat{A}$, we have
$$\hat{B}(\hat{C} + i\hat{A}) = \hat{B}\hat{C} + i\hat{B}\hat{A} = i\hbar\hat{A} + \hat{C}\hat{B} + i(\hat{A}\hat{B} - i\hbar\hat{C})$$
$$= (\hat{C} + i\hat{A})\hat{B} + \hbar(\hat{C} + i\hat{A}) = (\hat{C} + i\hat{A})(\hat{B} + \hbar),$$

$$(10.8)$$

as required.

Similarly

$$\hat{B}(\hat{C} - i\hat{A}) = \hat{B}\hat{C} - i\hat{B}\hat{A} = i\hbar\hat{A} + \hat{C}\hat{B} - i(\hat{A}\hat{B} - i\hbar\hat{C})$$
$$= (\hat{C} - i\hat{A})\hat{B} - \hbar(\hat{C} - i\hat{A}) = (\hat{C} - i\hat{A})(\hat{B} - \hbar),$$
(10.9)

as required.

These two relations show that

$$\left[\hat{B}, (\hat{C} \pm i\hat{A})\right] = \pm\hbar(\hat{C} \pm i\hat{A}). \qquad (10.10)$$

Problem 10.4

Show that

$$e^{\hat{A}}\hat{B}e^{-\hat{A}} = \hat{B} + \frac{1}{1!}[\hat{A}, \hat{B}] + \frac{1}{2!}[\hat{A}, [\hat{A}, \hat{B}]] + \frac{1}{3!}[\hat{A}, [\hat{A}, [\hat{A}, \hat{B}]]] + \dots$$
(10.11)

This formula shows that the calculation of complicated exponential-type operator functions can be simplified to the calculation of a series of commutators.

Solution

Expanding the exponents into the Taylor series

$$e^{\pm\hat{A}} = 1 \pm \frac{1}{1!}\hat{A} + \frac{1}{2!}\hat{A}^2 \pm \frac{1}{3!}\hat{A}^3 + \dots, \qquad (10.12)$$

we obtain

$$e^{\hat{A}}\hat{B}e^{-\hat{A}} = \left(1 + \frac{1}{1!}\hat{A} + \frac{1}{2!}\hat{A}^2 + \frac{1}{3!}\hat{A}^3 + \dots\right)$$
$$\hat{B}\left(1 - \frac{1}{1!}\hat{A} + \frac{1}{2!}\hat{A}^2 - \frac{1}{3!}\hat{A}^3 + \dots\right)$$
$$= \left(\hat{B} + \frac{1}{1!}\hat{A}\hat{B} + \frac{1}{2!}\hat{A}^2\hat{B} + \frac{1}{3!}\hat{A}^3\hat{B} + \dots\right)$$
$$\left(1 - \frac{1}{1!}\hat{A} + \frac{1}{2!}\hat{A}^2 - \frac{1}{3!}\hat{A}^3 + \dots\right)$$
$$= \hat{B} - \hat{B}\hat{A} + \frac{1}{2}\hat{B}\hat{A}^2 - \frac{1}{3!}\hat{B}\hat{A}^3 + \dots + \hat{A}\hat{B} - \hat{A}\hat{B}\hat{A} + \frac{1}{2}\hat{A}\hat{B}\hat{A}^2 + \dots$$

$$+\frac{1}{2}\hat{A}^2\hat{B} - \frac{1}{2}\hat{A}^2\hat{B}\hat{A} + \frac{1}{3!}\hat{A}^3\hat{B} + \dots$$

$$= \hat{B} + [\hat{A}, \hat{B}] + \frac{1}{2}\left(\hat{A}^2\hat{B} + \hat{B}\hat{A}^2 - 2\hat{A}\hat{B}\hat{A}\right)$$

$$+\frac{1}{3!}\left(\hat{A}^3\hat{B} - \hat{B}\hat{A}^3 + 3\hat{A}\hat{B}\hat{A}^2 - 3\hat{A}^2\hat{B}\hat{A}\right) + \dots$$

$$= \hat{B} + [\hat{A}, \hat{B}] + \frac{1}{2}\left[\hat{A}\left(\hat{A}\hat{B} - \hat{B}\hat{A}\right) - \left(\hat{A}\hat{B} - \hat{B}\hat{A}\right)\hat{A}\right]$$

$$+\frac{1}{3!}\left[\hat{A}\left(\hat{A}^2\hat{B} + \hat{B}\hat{A}^2\right) - \left(\hat{A}^2\hat{B} + \hat{B}\hat{A}^2\right)\hat{A} + 2\hat{A}\hat{B}\hat{A}^2 - 2\hat{A}^2\hat{B}\hat{A}\right]$$

$$= \hat{B} + [\hat{A}, \hat{B}] + \frac{1}{2}[\hat{A}, [\hat{A}, \hat{B}]] + \frac{1}{3!}[\hat{A}, [\hat{A}, [\hat{A}, \hat{B}]]] + \dots$$

$$= \hat{B} + \frac{1}{1!}[\hat{A}, \hat{B}] + \frac{1}{2!}[\hat{A}, [\hat{A}, \hat{B}]] + \frac{1}{3!}[\hat{A}, [\hat{A}, [\hat{A}, \hat{B}]]] + \dots$$

$$\tag{10.13}$$

Problem 10.5

Consider two arbitrary operators $\hat{A}$ and $\hat{B}$. If $\hat{A}$ commutes with their commutator $[\hat{A}, \hat{B}]$:

(a) Prove that for a positive integer n

$$[\hat{A}^n, \hat{B}] = n\hat{A}^{n-1}[\hat{A}, \hat{B}]. \tag{10.14}$$

(b) Apply the commutation relation (a) to the special case of $\hat{A} = \hat{x}$, $\hat{B} = \hat{p}_x$, and show that

$$[f(\hat{x}), \hat{p}_x] = i\hbar\frac{df}{dx}, \tag{10.15}$$

assuming that $f(\hat{x})$ can be expanded in a power series of the operator $\hat{x}$.

Solution (a)

We will prove this relation by induction. First, we check if this relation is true for $n = 1$

$$[\hat{A}^1, \hat{B}] = \hat{A}^0 [\hat{A}, \hat{B}] = [\hat{A}, \hat{B}]. \qquad (10.16)$$

Assume that the relation is true for $n = k$. We will prove that the relation is true for $n = k + 1$, i.e.,

$$[\hat{A}^{k+1}, \hat{B}] = (k+1)\hat{A}^k [\hat{A}, \hat{B}]. \qquad (10.17)$$

Consider the left-hand side of the relation. Since

$$\hat{A} [\hat{A}, \hat{B}] = [\hat{A}, \hat{B}] \hat{A}, \qquad (10.18)$$

we obtain

$$
\begin{aligned}
L &= [\hat{A}^{k+1}, \hat{B}] = \hat{A}^{k+1}\hat{B} - \hat{B}\hat{A}^{k+1} = \hat{A}^k(\hat{A}\hat{B} - \hat{B}\hat{A}) + \hat{A}^k\hat{B}\hat{A} - \hat{B}\hat{A}^{k+1} \\
&= \hat{A}^k [\hat{A}, \hat{B}] + (\hat{A}^k\hat{B} - \hat{B}\hat{A}^k)\hat{A} = \hat{A}^k [\hat{A}, \hat{B}] + [\hat{A}^k, \hat{B}] \hat{A} \\
&= \hat{A}^k [\hat{A}, \hat{B}] + k\hat{A}^{k-1} [\hat{A}, \hat{B}] \hat{A} \\
&= \hat{A}^k [\hat{A}, \hat{B}] + k\hat{A}^k [\hat{A}, \hat{B}] = (k+1)\hat{A}^k [\hat{A}, \hat{B}] = R, \text{ as required.}
\end{aligned}
$$
$$(10.19)$$

Solution (b)

Expanding $f(\hat{x})$ into the Taylor series

$$f(\hat{x}) = \sum_n \frac{1}{n!} \left(\frac{d^n f}{dx^n}\right)_{x=0} \hat{x}^n, \qquad (10.20)$$

we obtain

$$[f(\hat{x}), \hat{p}_x] = \sum_n \frac{1}{n!} \left(\frac{d^n f}{dx^n}\right)_{x=0} [\hat{x}^n, \hat{p}_x]. \qquad (10.21)$$

Now, applying the commutation relation from (a), we find

$$\sum_n \frac{1}{n!} \left(\frac{d^n f}{dx^n}\right)_{x=0} [\hat{x}^n, \hat{p}_x] = \sum_n \frac{1}{n!} \left(\frac{d^n f}{dx^n}\right)_{x=0} n\hat{x}^{n-1} [\hat{x}, \hat{p}_x]. \qquad (10.22)$$

Since

$$[\hat{x}, \hat{p}_x] = i\hbar, \qquad (10.23)$$

we obtain

$$\sum_n \frac{1}{n!} \left(\frac{d^n f}{dx^n}\right)_{x=0} n\hat{x}^{n-1} [\hat{x}, \hat{p}_x] = \sum_n \frac{1}{n!} \left(\frac{d^n f}{dx^n}\right)_{x=0} n\hat{x}^{n-1} i\hbar$$

$$= i\hbar \frac{d}{dx} \left(\sum_n \frac{1}{n!} \left(\frac{d^{n-1} f}{dx^{n-1}}\right)_{x=0} n\hat{x}^{n-1}\right)$$

$$= i\hbar \frac{d}{dx} \left(\sum_n \frac{1}{(n-1)!} \left(\frac{d^{n-1} f}{dx^{n-1}}\right)_{x=0} \hat{x}^{n-1}\right). \tag{10.24}$$

By substituting $n - 1 = k$, we get

$$i\hbar \frac{d}{dx} \left(\sum_n \frac{1}{(n-1)!} \left(\frac{d^{n-1} f}{dx^{n-1}}\right)_{x=0} \hat{x}^{n-1}\right)$$

$$= i\hbar \frac{d}{dx} \left(\sum_k \frac{1}{k!} \left(\frac{d^k f}{dx^k}\right)_{x=0} \hat{x}^k\right) = i\hbar \frac{df}{dx}. \tag{10.25}$$

Problem 10.7

Determine if the function $\phi = e^{ax} \sin x$, where a is a real constant, is an eigenfunction of the operator d/dx and d^2/dx^2. If it is, determine any eigenvalue.

Solution

The function ϕ is an eigenfunction of an operator $\hat{A}$ if

$$\hat{A}\phi = \alpha\phi, \tag{10.26}$$

where α is an eigenvalue.

We wish to find if the function ϕ is an eigenfunction of the operators d/dx and d^2/dx^2.

Calculate

$$\frac{d\phi}{dx} \quad \text{and} \quad \frac{d^2\phi}{dx^2}. \tag{10.27}$$

Since

$$\frac{d\phi}{dx} = ae^{ax} \sin x + e^{ax} \cos x = a\phi + e^{ax} \cos x, \tag{10.28}$$

we see that

$$\frac{d\phi}{dx} \neq \alpha\phi \qquad \text{for all } a. \qquad (10.29)$$

From this result we see that ϕ is not an eigenfunction of the operator d/dx.

Consider now the operator d^2/dx^2. Since

$$\frac{d^2\phi}{dx^2} = a^2 e^{ax} \sin x + ae^{ax} \cos x + ae^{ax} \cos x - e^{ax} \sin x$$

$$= (a^2 - 1)\phi + 2ae^{ax} \cos x, \qquad (10.30)$$

we can identify that for $a = 0$, ϕ is an eigenfunction of d^2/dx^2 with an eigenvalue $\alpha = -1$.

Problem 10.9

Calculate the expectation value of the x coordinate of a particle in the energy state E_n of a one-dimensional box.

Solution

The expectation value of x is

$$\langle x \rangle = \int \phi^*(x) x \phi(x) dx, \qquad (10.31)$$

where $\phi(x)$ is the wave function of the particle.

Since

$$\phi(x) = \phi_n(x) = \sqrt{\frac{2}{a}} \sin\left(\frac{n\pi}{a}x\right), \qquad \text{for} \quad 0 \leq x \leq a, \quad (10.32)$$

and $\phi(x)$ is zero for $x < 0$ and $x > a$, we obtain

$$\langle x \rangle = \int \phi^*(x) x \phi(x) dx = \frac{2}{a} \int_0^a dx \, x \sin^2 \left(\frac{n\pi}{a} x \right)$$

$$= \frac{1}{a} \int_0^a dx \, x \left[1 - \cos \left(\frac{2n\pi}{a} x \right) \right]$$

$$= \frac{1}{a} \int_0^a dx \left[x - x \cos \left(\frac{2n\pi}{a} x \right) \right]$$

$$= \frac{1}{a} \int_0^a dx \, x - \frac{1}{a} \int_0^a dx \, x \cos \left(\frac{2n\pi}{a} x \right)$$

$$= \frac{1}{a} \frac{1}{2} x^2 \Big|_0^a - \frac{1}{a} \left\{ -\frac{a^2}{(2n\pi)^2} \left[\cos \left(\frac{2n\pi}{a} x \right) + \frac{2n\pi x}{a} \sin \left(\frac{2n\pi}{a} x \right) \right]_0^a \right\}.$$
(10.33)

Since $\cos(2n\pi) = \cos 0 = 1$ and $\sin(2n\pi) = \sin 0 = 0$, it follows that

$$\langle x \rangle = \frac{1}{2} a, \tag{10.34}$$

independent of n. Physically, this value results from the fact that the wave function of the particle is symmetric about $x = a/2$ for all n. Note, the expectation value is not equal to the most probable value, which is given by $|\phi(x)|^2$.

Problem 10.11

For a particle in an infinite square well potential represented by the position $\hat{x}$ and momentum $\hat{p}_x$ operators, check the uncertainty principle $\Delta x \Delta p_x \geq \hbar/2$ for $n = 1$, where $\Delta x = \sqrt{\langle \hat{x}^2 \rangle - \langle \hat{x} \rangle^2}$ and $\Delta p_x = \sqrt{\langle \hat{p}_x^2 \rangle - \langle \hat{p}_x \rangle^2}$.

Solution

For $n = 1$ the wave function of the particle in the infinite square well potential is given by

$$\phi_1(x) = \sqrt{\frac{2}{a}} e^{\frac{ika}{2}} \sin \left[\frac{\pi}{a} \left(x - \frac{a}{2} \right) \right], \tag{10.35}$$

which can be written as

$$\phi_1(x) = \sqrt{\frac{2}{a}} e^{\frac{ika}{2}} \cos\left(\frac{\pi}{a}x\right). \tag{10.36}$$

First, calculate the average $\langle x \rangle$. From the definition of the expectation value (average), we have

$$\langle x \rangle = \int \phi_1^*(x) x \phi_1(x) dx = \frac{2}{a} \int_{-a/2}^{a/2} dx\, x \cos^2\left(\frac{\pi}{a}x\right). \tag{10.37}$$

The integral is zero, as the function under the integral is an odd function and the integral is taken over a range that is centered about the origin. Thus, $\langle x \rangle = 0$.

We now calculate $\langle x^2 \rangle$. From the definition of the expectation value, we get

$$\langle x^2 \rangle = \int \phi_1^*(x) x^2 \phi_1(x) dx = \frac{2}{a} \int_{-a/2}^{a/2} dx\, x^2 \cos^2\left(\frac{\pi}{a}x\right)$$

$$= \frac{4}{a} \int_0^{a/2} dx\, x^2 \cos^2\left(\frac{\pi}{a}x\right) = \frac{2}{a} \int_0^{a/2} dx\, x^2 \left[1 + \cos\left(\frac{2\pi}{a}x\right)\right]$$

$$= \frac{2}{a} \left[\int_0^{a/2} dx\, x^2 + \int_0^{a/2} dx\, x^2 \cos\left(\frac{2\pi}{a}x\right)\right]. \tag{10.38}$$

In the second integral, we replace $2\alpha x$ by z, where $\alpha = \pi/a$, and then integrating by parts, we obtain

$$\langle x^2 \rangle = \frac{2}{a} \left[\int_0^{a/2} dx\, x^2 + \frac{1}{8\alpha^3} \int_0^{\pi} dz\, z^2 \cos z\right]$$

$$= \frac{2}{a} \left[\frac{x^3}{3}\bigg|_0^{a/2} + \frac{1}{8\alpha^3} \left(2z \cos z + (z^2 - 2)\sin z\right)\bigg|_0^{\pi}\right]$$

$$= \frac{2}{a} \left[\frac{a^3}{24} + \frac{1}{8\alpha^3}(-2\pi)\right] = \frac{2}{a}\left(\frac{a^3}{24} - \frac{\pi}{4}\frac{a^3}{\pi^3}\right) = \frac{a^2}{2\pi^2}\left(\frac{\pi^2}{6} - 1\right). \tag{10.39}$$

Hence, the variance Δx is

$$\Delta x = \sqrt{\langle \hat{x}^2 \rangle - \langle \hat{x} \rangle^2} = \frac{a}{2\pi}\sqrt{2\left(\frac{\pi^2}{6} - 1\right)}. \tag{10.40}$$

Calculate now $\langle \hat{p}_x \rangle$. Since

$$\hat{p}_x = -i\hbar \frac{d}{dx}, \tag{10.41}$$

and

$$\hat{p}_x \phi_1(x) = -i\hbar \frac{d\phi_1(x)}{dx} = -i\hbar \sqrt{\frac{2}{a}} e^{\frac{ika}{2}} \alpha \cos(\alpha x), \quad (10.42)$$

we obtain

$$\langle \hat{p}_x \rangle = \int \phi_1^*(x) \hat{p}_x \phi_1(x) dx = -i\hbar \frac{2}{a} \int_{-a/2}^{a/2} dx \cos \alpha x \, (-\alpha \sin \alpha x)$$

$$= \frac{1}{2} i\hbar\alpha \int_{-a/2}^{a/2} dx \, \sin(2\alpha x). \quad (10.43)$$

Since the function under the integral is an odd function and the integral is taken over a range which is centered about the origin, the integral is zero. Thus,

$$\langle \hat{p}_x \rangle = 0. \quad (10.44)$$

Calculate $\langle \hat{p}_x^2 \rangle$:

$$\langle \hat{p}_x^2 \rangle = \int \phi_1^*(x) \hat{p}_x^2 \phi_1(x) dx = -\hbar^2 \frac{2}{a} \int_{-a/2}^{a/2} dx \, \cos(\alpha x) \frac{d^2}{dx^2} \cos(\alpha x)$$

$$= \hbar^2 \alpha^2 \frac{2}{a} \int_{-a/2}^{a/2} dx \, \cos^2(\alpha x). \quad (10.45)$$

However,

$$\int_{-a/2}^{a/2} dx \, \cos^2(\alpha x) = \frac{1}{2} \int_{-a/2}^{a/2} dx \, [1 + \cos(2\alpha x)] = \frac{a}{2}, \quad (10.46)$$

and therefore

$$\langle \hat{p}_x^2 \rangle = \hbar^2 \alpha^2. \quad (10.47)$$

Hence

$$\Delta p_x = \alpha\hbar = \frac{\pi\hbar}{a}. \quad (10.48)$$

Combining this result with that for Δx gives

$$\Delta x \Delta p_x = \frac{a}{2\pi} \sqrt{2\left(\frac{\pi^2}{6} - 1\right)} \frac{\pi\hbar}{a} = \frac{\hbar}{2} \sqrt{2\left(\frac{\pi^2}{6} - 1\right)}. \quad (10.49)$$

But, since

$$\sqrt{2\left(\frac{\pi^2}{6} - 1\right)} > 1, \quad (10.50)$$

it follows that

$$\Delta x \Delta p_x > \frac{\hbar}{2}, \quad (10.51)$$

i.e., the uncertainty principle is satisfied.

Problem 10.12

The expectation value of an arbitrary operator $\hat{A}$ in the state $\phi(x)$ is given by

$$\langle \hat{A} \rangle = \int \phi^*(x)\hat{A}\phi(x)dx. \tag{10.52}$$

(a) Calculate expectation values (i) $\langle \hat{x}\hat{p}_x \rangle$, (ii) $\langle \hat{p}_x\hat{x} \rangle$, and (iii) $(\langle \hat{x}\hat{p}_x \rangle + \langle \hat{p}_x\hat{x} \rangle)/2$ of the product of position ($\hat{x} = x$) and momentum ($\hat{p}_x = -i\hbar\frac{d}{dx}$) operators of a particle represented by the wave function

$$\phi(x) = \sqrt{\frac{2}{a}}\sin\left(\frac{\pi x}{a}\right), \tag{10.53}$$

where $0 \le x \le a$.

(b) The operators $\hat{x}$ and $\hat{p}_x$ are Hermitian. Which of the products (i) $\hat{x}\hat{p}_x$, (ii) $\hat{p}_x\hat{x}$, and (iii) $(\hat{x}\hat{p}_x + \hat{p}_x\hat{x})/2$ are Hermitian?

(c) Explain, which of the results of **(a)** are acceptable as the expectation values of physical quantities.

Solution (a)

(i) Consider the expectation value of $\hat{x}\hat{p}_x$ in the state $\phi(x)$:

$$\langle \hat{x}\hat{p}_x \rangle = \int \phi^*(x)\hat{x}\hat{p}_x\phi(x)dx$$

$$= -i\hbar\frac{2}{a}\int_0^a dx \sin\left(\frac{\pi x}{a}\right) x \frac{d}{dx}\sin\left(\frac{\pi x}{a}\right)$$

$$= -i\frac{2\pi\hbar}{a^2}\int_0^a dx \sin\left(\frac{\pi x}{a}\right) x \cos\left(\frac{\pi x}{a}\right)$$

$$= -i\frac{\pi\hbar}{a^2}\int_0^a dx\, x \sin\left(\frac{2\pi x}{a}\right). \tag{10.54}$$

By making the substitution $\alpha = 2\pi x/a$, the previous integral then becomes

$$\langle \hat{x}\hat{p}_x \rangle = -i\frac{\pi\hbar}{a^2}\int_0^a dx\, x \sin\left(\frac{2\pi x}{a}\right) = -i\frac{\hbar}{4\pi}\int_0^{2\pi} d\alpha\, \alpha \sin\alpha. \tag{10.55}$$

The integration over α is readily performed by parts to give

$$\langle \hat{x}\,\hat{p}_x \rangle = -i\frac{\hbar}{4\pi}\left[-\alpha\cos\alpha|_0^{2\pi} + \int_0^{2\pi} d\alpha\,\cos\alpha\right]$$

$$= -i\frac{\hbar}{4\pi}\left[-2\pi + \sin\alpha|_0^{2\pi}\right] = \frac{1}{2}i\hbar. \qquad (10.56)$$

The expectation value of $\hat{x}\,\hat{p}_x$ is a complex number. It means that $\hat{x}\,\hat{p}_x$ is not, Hermitian operator. We will show it more explicitly in part (b).

(ii) Calculate now the expectation value of $\hat{p}_x\hat{x}$ in the state $\phi(x)$. Calculations similar to those performed in (i) give

$$\langle \hat{p}_x\hat{x} \rangle = \int \phi^*(x)\hat{p}_x\hat{x}\phi(x)dx = -i\hbar\frac{2}{a}\int_0^a dx\,\sin\left(\frac{\pi x}{a}\right)\frac{d}{dx}\left[x\sin\left(\frac{\pi x}{a}\right)\right]$$

$$= -i\frac{2\hbar}{a}\int_0^a dx\,\sin\left(\frac{\pi x}{a}\right)\left[\sin\left(\frac{\pi x}{a}\right) + \frac{\pi}{a}x\cos\left(\frac{\pi x}{a}\right)\right]$$

$$= -i\frac{2\hbar}{a}\int_0^a dx\,\sin^2\left(\frac{\pi x}{a}\right) - i\frac{2\pi\hbar}{a^2}\int_0^a dx\,x\sin\left(\frac{\pi x}{a}\right)\cos\left(\frac{\pi x}{a}\right)$$

$$= -i\frac{2\hbar}{a}\int_0^a dx\,\sin^2\left(\frac{\pi x}{a}\right) - i\frac{\pi\hbar}{a^2}\int_0^a dx\,x\sin\left(\frac{2\pi x}{a}\right)$$

$$= -i\frac{\hbar}{a}\int_0^a dx\,\left[1 - \cos\left(\frac{2\pi x}{a}\right)\right] - i\frac{\pi\hbar}{a^2}\int_0^a dx\,x\sin\left(\frac{2\pi x}{a}\right)$$

$$= -i\hbar + \frac{1}{2}i\hbar = -\frac{1}{2}i\hbar.$$

$$(10.57)$$

(iii) Since the expectation value is additive, tht expectation value of $(\hat{x}\,\hat{p}_x + \hat{p}_x\hat{x})/2$ is obtained simply by adding the results (i) and (ii):

$$\left\langle \frac{1}{2}(\hat{x}\,\hat{p}_x + \hat{p}_x\hat{x}) \right\rangle = \frac{1}{2}\left(\langle \hat{x}\,\hat{p}_x \rangle + \langle \hat{p}_x\hat{x} \rangle\right) = \frac{1}{2}\left(\frac{1}{2}i\hbar - \frac{1}{2}i\hbar\right) = 0.$$

$$(10.58)$$

Solution (b)

From the definition, an operator $\hat{A}$ is Hermitian if $\hat{A}^\dagger = \hat{A}$.

(i) Consider a Hermitian conjugate of $\hat{x}\hat{p}_x$. Since $\hat{x}$ and $\hat{p}_x$ are Hermitian operators and they do not commute, we get

$$(\hat{x}\,\hat{p}_x)^\dagger = \hat{p}_x^\dagger\hat{x}^\dagger = \hat{p}_x\hat{x} \neq \hat{x}\,\hat{p}_x. \qquad (10.59)$$

Hence, $\hat{x}\hat{p}_x$ is not Hermitian. It is interesting that a product of two Hermitian operators does not have to be Hermitian.

(ii) Consider $\hat{p}_x\hat{x}$. A Hermitian conjugate of $\hat{p}_x\hat{x}$ is

$$(\hat{p}_x\hat{x})^\dagger = \hat{x}^\dagger \hat{p}_x^\dagger = \hat{x}\hat{p}_x \neq \hat{p}_x\hat{x}. \qquad (10.60)$$

Hence, $\hat{p}_x\hat{x}$ is not Hermitian.

(iii) Taking a Hermitian conjugate of $(\hat{x}\hat{p}_x + \hat{p}_x\hat{x})/2$ and using the results of (i) and (ii), we find

$$[(\hat{x}\hat{p}_x + \hat{p}_x\hat{x})/2]^\dagger = (\hat{p}_x^\dagger\hat{x}^\dagger + \hat{x}^\dagger\hat{p}_x^\dagger)/2 = (\hat{p}_x\hat{x} + \hat{x}\hat{p}_x)/2$$
$$= (\hat{x}\hat{p}_x + \hat{p}_x\hat{x})/2. \qquad (10.61)$$

Hence, $(\hat{x}\hat{p}_x + \hat{p}_x\hat{x})/2$ is Hermitian.

Solution (c)

Physical quantities are represented by Hermitian operators. Hence, only (iii) $\langle(\hat{x}\hat{p}_x + \hat{p}_x\hat{x})/2\rangle$ is acceptable as the expectation value of a physical quantity.

Chapter 11

Dirac Bra-Ket Notation

Problem 11.1

Useful application of the completeness relation

As we have mentioned in the chapter, the completeness relation is very useful in calculations involving operators and state vectors. Consider the following example.

Let A_{il} and B_{lj} be matrix elements of two arbitrary operators $\hat{A}$ and $\hat{B}$ in a basis of orthonormal vectors. Show, using the completeness relation, that matrix elements of the product operator $\hat{A}\hat{B}$ in the same orthonormal basis can be found from the multiplication of the matrix elements A_{il} and B_{lj} as

$$\left(\hat{A}\hat{B}\right)_{ij} = \sum_{l=1}^{n} A_{il} B_{lj}. \tag{11.1}$$

Solution

The completeness relation for a set of orthonormal states $|\phi_i\rangle$

$$\sum_{i=1}^{n} |\phi_i\rangle\langle\phi_i| = 1, \tag{11.2}$$

Problems and Solutions in Quantum Physics
Zbigniew Ficek
Copyright © 2016 Pan Stanford Publishing Pte. Ltd.
ISBN 978-981-4669-36-8 (Hardcover), 978-981-4669-37-5 (eBook)
www.panstanford.com

can be used to represent an arbitrary operator in terms of the orthonormal states. Consider an operator $\hat{C}$, which is a product of two operators $\hat{A}$ and $\hat{B}$:

$$\hat{C} = \hat{A}\hat{B}. \tag{11.3}$$

Multiplying the operator $\hat{C}$ by the unity given by Eq. (11.2) both on the right and the left, we then obtain the operator in terms of the projection operators $|\phi_i\rangle\langle\phi_i|$ as

$$\hat{C} = \left(\sum_i |\phi_i\rangle\langle\phi_i|\right) \hat{C} \left(\sum_j |\phi_j\rangle\langle\phi_j|\right)$$

$$= \sum_{i,j} \langle\phi_i|\hat{C}|\phi_j\rangle|\phi_i\rangle\langle\phi_j| = \sum_{i,j} c_{ij}|\phi_i\rangle\langle\phi_j|, \tag{11.4}$$

where $c_{ij} = \langle\phi_i|\hat{C}|\phi_j\rangle$ are matrix elements of the operator $\hat{C}$ in the basis of the states $|\phi_i\rangle$. Since $\hat{C} = \hat{A}\hat{B}$, we can write $c_{ij} = \langle\phi_i|\hat{A}\hat{B}|\phi_j\rangle \equiv \left(\hat{A}\hat{B}\right)_{ij}$.

Applying the completeness relation in between the operators $\hat{A}$ and $\hat{B}$, we get

$$\left(\hat{A}\hat{B}\right)_{ij} = \langle\phi_i|\hat{A}\hat{B}|\phi_j\rangle = \langle\phi_i|\hat{A}\left(\sum_l |\phi_l\rangle\langle\phi_l|\right)\hat{B}|\phi_j\rangle$$

$$= \sum_{l=1}^{n} \langle\phi_i|\hat{A}|\phi_l\rangle\langle\phi_l|\hat{B}|\phi_j\rangle = \sum_{l=1}^{n} A_{il}B_{lj}. \tag{11.5}$$

Thus, matrix elements of the product operator $\hat{A}\hat{B}$ in an orthonormal basis are equal to the product of the matrix elements A_{il} and B_{lj} of the operators $\hat{A}$ and $\hat{B}$ represented in the same basis.

Problem 11.2

Eigenvalues of the projection operator

Show that the eigenvalues of the projection operator P_{nn} are 0 or 1.

Solution

Suppose that a state $|m\rangle$ is an eigenstate of the projection operator with an eigenvalue α:

$$P_{nn}|m\rangle = \alpha|m\rangle. \tag{11.6}$$

Since $P_{nn} = |n\rangle\langle n|$, we have

$$P_{nn}|m\rangle = |n\rangle\langle n|m\rangle = |n\rangle\delta_{nm} = \alpha|m\rangle. \tag{11.7}$$

Multiplying both sides from the left by $\langle m|$, we get

$$\langle m|n\rangle\delta_{nm} = \delta_{nm}^2 = \alpha. \tag{11.8}$$

For $n = m$, $\delta_{nm} = 1$, whereas for $n \neq m$, $\delta_{nm} = 0$. Thus, $\alpha = 0, 1$.

Problem 11.3

Sum of two diagonal projection operators

Let P_{nn} and P_{mm} be diagonal projection operators. Show that the sum $P_{nn} + P_{mm}$ is a diagonal projection operator if and only if $P_{nn}P_{mm} = 0$.

Solution

A diagonal projection operator has the property $P_{kk}^2 = P_{kk}$. Suppose that $P_{kk} = P_{nn} + P_{mm}$. Then

$$P_{kk}^2 = (P_{nn}+P_{mm})(P_{nn}+P_{mm}) = P_{nn}P_{nn}+P_{nn}P_{mm}+P_{mm}P_{nn}+P_{mm}P_{mm}. \tag{11.9}$$

Since $P_{nn}P_{nn} = P_{nn}$ and $P_{mm}P_{mm} = P_{mm}$, we have

$$P_{kk}^2 = P_{nn} + P_{nn}P_{mm} + P_{mm}P_{nn} + P_{mm}. \tag{11.10}$$

Hence, $P_{kk}^2 = P_{kk}$ only if $P_{nn}P_{mm} = 0$.

Chapter 12

Matrix Representations

Problem 12.1

Eigenvalues and eigenvectors of Hermitian operators.

(a) Consider two Hermitian operators $\hat{A}$ and $\hat{B}$ that have the same complete set of eigenfunctions ϕ_n. Show that the operators commute.

(b) Suppose two Hermitian operators have the matrix representation:

$$\hat{A} = \begin{bmatrix} a & 0 & 0 \\ 0 & -a & 0 \\ 0 & 0 & -a \end{bmatrix}, \ \hat{B} = \begin{bmatrix} b & 0 & 0 \\ 0 & 0 & ib \\ 0 & -ib & 0 \end{bmatrix}, \qquad (12.1)$$

where a and b are real numbers.

(i) Calculate the eigenvalues of $\hat{A}$ and $\hat{B}$.
(ii) Show that $\hat{A}$ and $\hat{B}$ commute.
(iii) Determine a complete set of common eigenfunctions.

Problems and Solutions in Quantum Physics
Zbigniew Ficek
Copyright © 2016 Pan Stanford Publishing Pte. Ltd.
ISBN 978-981-4669-36-8 (Hardcover), 978-981-4669-37-5 (eBook)
www.panstanford.com

Solution (a)

Assume that ϕ_n are eigenfunctions of $\hat{A}$ with corresponding eigenvalues α_n, and eigenfunctions of $\hat{B}$ with corresponding eigenvalues β_n. Then

$$\hat{A}\hat{B}\phi_n = \hat{A}\left(\beta_n\phi_n\right) = \beta_n\hat{A}\phi_n = \beta_n\alpha_n\phi_n. \tag{12.2}$$

Similarly

$$\hat{B}\hat{A}\phi_n = \hat{B}\left(\alpha_n\phi_n\right) = \alpha_n\hat{B}\phi_n = \alpha_n\beta_n\phi_n. \tag{12.3}$$

Hence

$$\hat{A}\hat{B}\phi_n - \hat{B}\hat{A}\phi_n = \left[\hat{A}, \hat{B}\right]\phi_n = \left(\beta_n\alpha_n - \alpha_n\beta_n\right)\phi_n = 0. \tag{12.4}$$

Consequently, $\left[\hat{A}, \hat{B}\right] = 0$.

Solution (b)

(i) Consider an eigenvalue equation for $\hat{A}$:

$$\begin{pmatrix} a & 0 & 0 \\ 0 & -a & 0 \\ 0 & 0 & -a \end{pmatrix} \begin{pmatrix} c_1 \\ c_2 \\ c_3 \end{pmatrix} = \lambda \begin{pmatrix} c_1 \\ c_2 \\ c_3 \end{pmatrix}, \tag{12.5}$$

which can be written as

$$\begin{pmatrix} a - \lambda & 0 & 0 \\ 0 & -a - \lambda & 0 \\ 0 & 0 & -a - \lambda \end{pmatrix} \begin{pmatrix} c_1 \\ c_2 \\ c_3 \end{pmatrix} = 0. \tag{12.6}$$

This equation has nonzero solutions when the determinant of the matrix is zero, i.e., when

$$\begin{vmatrix} a - \lambda & 0 & 0 \\ 0 & -a - \lambda & 0 \\ 0 & 0 & -a - \lambda \end{vmatrix} = 0. \tag{12.7}$$

From this we find a cubic equation

$$(\lambda - a)(\lambda + a)^2 = 0. \tag{12.8}$$

The roots of the cubic equation are

$$\lambda_1 = a, \quad \lambda_2 = -a, \quad \lambda_3 = -a. \tag{12.9}$$

Thus, the eigenvalues of the matrix $\hat{A}$ are $\lambda_1 = a$ and $\lambda_{2,3} = -a$. Note that the eigenvalues λ_2 and λ_3 are degenerated.

Consider now an eigenvalue equation for $\hat{B}$:

$$\begin{pmatrix} b & 0 & 0 \\ 0 & 0 & ib \\ 0 & -ib & 0 \end{pmatrix} \begin{pmatrix} c_1 \\ c_2 \\ c_3 \end{pmatrix} = \lambda \begin{pmatrix} c_1 \\ c_2 \\ c_3 \end{pmatrix}. \tag{12.10}$$

The equation can be written as

$$\begin{pmatrix} b-\lambda & 0 & 0 \\ 0 & -\lambda & ib \\ 0 & -ib & -\lambda \end{pmatrix} \begin{pmatrix} c_1 \\ c_2 \\ c_3 \end{pmatrix} = 0. \tag{12.11}$$

This equation has nonzero solutions when the determinant of the matrix is zero, i.e., when

$$\begin{vmatrix} b-\lambda & 0 & 0 \\ 0 & -\lambda & ib \\ 0 & -ib & -\lambda \end{vmatrix} = 0. \tag{12.12}$$

From this we find a cubic equation

$$(\lambda^2 - b^2)(\lambda - b) = 0. \tag{12.13}$$

The roots of the cubic equation are

$$\lambda_1 = b, \quad \lambda_2 = -b, \quad \lambda_3 = b. \tag{12.14}$$

Thus, the eigenvalues of the matrix $\hat{B}$ are $\lambda_{1,3} = b$ and $\lambda_2 = -b$. Similar to the operator $\hat{A}$, two eigenvalues λ_1 and λ_2 are degenerated.

(ii) Consider the commutator $[\hat{A}, \hat{B}] = \hat{A}\hat{B} - \hat{B}\hat{A}$. First, we calculate $\hat{A}\hat{B}$ and find

$$\hat{A}\hat{B} = \begin{pmatrix} a & 0 & 0 \\ 0 & -a & 0 \\ 0 & 0 & -a \end{pmatrix} \begin{pmatrix} b & 0 & 0 \\ 0 & 0 & ib \\ 0 & -ib & 0 \end{pmatrix} = \begin{pmatrix} ab & 0 & 0 \\ 0 & 0 & -iab \\ 0 & iab & 0 \end{pmatrix}. \tag{12.15}$$

Next, we calculate $\hat{B}\hat{A}$ and find

$$\hat{B}\hat{A} = \begin{pmatrix} b & 0 & 0 \\ 0 & 0 & ib \\ 0 & -ib & 0 \end{pmatrix} \begin{pmatrix} a & 0 & 0 \\ 0 & -a & 0 \\ 0 & 0 & -a \end{pmatrix} = \begin{pmatrix} ab & 0 & 0 \\ 0 & 0 & -iab \\ 0 & iab & 0 \end{pmatrix}. \tag{12.16}$$

Hence $\hat{A}\hat{B} = \hat{B}\hat{A}$. Thus, $\hat{A}$ and $\hat{B}$ commute.

(iii) Since the operators $\hat{A}$ and $\hat{B}$ commute, they have common eigenfunctions. Therefore, it is enough to find the eigenfunctions of either $\hat{A}$ or $\hat{B}$. Lets find the eigenfunctions of $\hat{A}$.

Notice that the matrix $\hat{A}$ is diagonal. This means that the basis states ϕ_1, ϕ_2, ϕ_3 are eigenstates (eigenfunctions) of the operator $\hat{A}$.

We have showed in (ii) that the operators $\hat{A}$ and $\hat{B}$ commute, so they have common eigenfunctions. It means that ϕ_1, ϕ_2, ϕ_3 are also the eigenfunctions of the operator $\hat{B}$. One could question this statement: Matrix $\hat{B}$ is not diagonal: then how could ϕ_1, ϕ_2, ϕ_3 be the eigenstates of matrix $\hat{B}$?

The answer is in the fact that matrix $\hat{B}$ has two degenerate eigenvalues $\lambda_2 = \lambda_3 = b$. If two eigenfunctions ϕ_2 and ϕ_3 of an operator $\hat{B}$ are degenerated, then not only ϕ_2 and ϕ_3 are the eigenfunctions of $\hat{B}$, but also any linear combination of ϕ_2 and ϕ_3 is an eigenfunction of $\hat{B}$. To prove this let us consider the eigenvalue equation for matrix $\hat{B}$:

$$\begin{pmatrix} b & 0 & 0 \\ 0 & 0 & ib \\ 0 & -ib & 0 \end{pmatrix} \begin{pmatrix} c_1 \\ c_2 \\ c_3 \end{pmatrix} = \lambda \begin{pmatrix} c_1 \\ c_2 \\ c_3 \end{pmatrix}. \tag{12.17}$$

For $\lambda = b$, the eigenvalue equation is of the form

$$\begin{pmatrix} b & 0 & 0 \\ 0 & 0 & ib \\ 0 & -ib & 0 \end{pmatrix} \begin{pmatrix} c_1 \\ c_2 \\ c_3 \end{pmatrix} = b \begin{pmatrix} c_1 \\ c_2 \\ c_3 \end{pmatrix}. \tag{12.18}$$

from which we find that

$$bc_1 = bc_1 \quad \text{and} \quad ibc_3 = bc_2. \tag{12.19}$$

This means that ϕ_1 is an eigenfunction of $\hat{B}$ with the eigenvalue b, and a linear superposition

$$\phi_b = ic_3\phi_2 + c_3\phi_3, \tag{12.20}$$

is also an eigenfunction of the operator $\hat{B}$ with the eigenvalue b.

From the normalization condition, we find that $c_3 = 1/\sqrt{2}$, and then the normalized eigenfunction ϕ_b is of the form

$$\phi_b = \frac{1}{\sqrt{2}} (i\phi_2 + \phi_3). \tag{12.21}$$

This clearly shows that in the case of degenerate eigenvalues of an operator, not only ϕ_1, ϕ_2, ... are eigenfunctions of the operator, but an arbitrary combination of ϕ_1, ϕ_2, ... is also an eigenfunction of the operator.

Problem 12.2

The Rabi problem illustrates what are the energy states of an atom driven by an external coherent (laser) field.

A laser field of frequency ω_L drives a transition in an atom between two atomic energy states $|1\rangle$ and $|2\rangle$. The states are separated by the frequency ω_0. The Hamiltonian of the system in the bases of the atomic states is given by the matrix

$$\hat{H} = \hbar \begin{pmatrix} -\frac{1}{2}\Delta & \Omega \\ \Omega & \frac{1}{2}\Delta \end{pmatrix}, \tag{12.22}$$

where $\Delta = \omega_L - \omega_0$ is the detuning of the laser frequency from the atomic transition frequency, and Ω is the Rabi frequency that describes the strength of the laser field acting on the atom.

Find the energies and energy states of the system, the so-called dressed states, which are, respectively, eigenvalues and eigenvectors of the Hamiltonian $\hat{H}$.

Solution

In general, the energy state of the system is a linear superposition of the energy states $|1\rangle$ and $|2\rangle$:

$$|\Psi\rangle = c_1|1\rangle + c_2|2\rangle, \tag{12.23}$$

where c_1 and c_2 are unknown amplitudes, which are to be determined solving an eigenvalue equation for the Hamiltonian $\hat{H}$. The eigenvalue equation for $\hat{H}$ is

$$\begin{pmatrix} -\frac{1}{2}\Delta & \Omega \\ \Omega & \frac{1}{2}\Delta \end{pmatrix} \begin{pmatrix} c_1 \\ c_2 \end{pmatrix} = \lambda \begin{pmatrix} c_1 \\ c_2 \end{pmatrix}, \tag{12.24}$$

which can be written as

$$\begin{pmatrix} -\frac{1}{2}\Delta - \lambda & \Omega \\ \Omega & \frac{1}{2}\Delta - \lambda \end{pmatrix} \begin{pmatrix} c_1 \\ c_2 \end{pmatrix} = 0. \tag{12.25}$$

This equation has nonzero solutions when the determinant of the matrix is zero, i.e., when

$$\begin{vmatrix} -\frac{1}{2}\Delta - \lambda & \Omega \\ \Omega & \frac{1}{2}\Delta - \lambda \end{vmatrix} = 0. \tag{12.26}$$

From this we find a quadratic equation

$$\lambda^2 - \frac{1}{4}\Delta^2 - \Omega^2 = 0, \tag{12.27}$$

whose roots are

$$\lambda_{1,2} = \pm\sqrt{\frac{1}{4}\Delta^2 + \Omega^2} \equiv \pm\tilde{\Omega}. \tag{12.28}$$

The eigenvalues $\lambda_{1,2}$ of the Hamiltonian are the energies of the system.

In order to find the energy states of the system, we have to find the eigenvectors of the Hamiltonian corresponding to the two eigenvalues, $\lambda_1 = \tilde{\Omega}$ and $\lambda_2 = -\tilde{\Omega}$.

For $\lambda_1 = \tilde{\Omega}$, the eigenvalue equation is of the form

$$\begin{pmatrix} -\frac{1}{2}\Delta & \Omega \\ \Omega & \frac{1}{2}\Delta \end{pmatrix}\begin{pmatrix} c_1 \\ c_2 \end{pmatrix} = \tilde{\Omega}\begin{pmatrix} c_1 \\ c_2 \end{pmatrix}, \tag{12.29}$$

from which we find an equation relating the coefficients c_1 and c_2:

$$-\frac{1}{2}\Delta c_1 + \Omega c_2 = \tilde{\Omega}c_1. \tag{12.30}$$

From this relation, we find

$$c_1 = \frac{\Omega}{\tilde{\Omega} + \frac{1}{2}\Delta}c_2. \tag{12.31}$$

Thus, the energy state of the system corresponding to the energy $\lambda_1 = \tilde{\Omega}$ is

$$|\Psi_1\rangle = \frac{\Omega}{\tilde{\Omega} + \frac{1}{2}\Delta}c_2|1\rangle + c_2|2\rangle. \tag{12.32}$$

The undetermined coefficient c_2 is found from the normalization condition $\langle\Psi_1|\Psi_1\rangle = 1$, which gives

$$c_2 = \sqrt{\frac{\tilde{\Omega} + \frac{1}{2}\Delta}{2\tilde{\Omega}}}. \tag{12.33}$$

Hence, the state $|\Psi_1\rangle$ takes the form

$$|\Psi_1\rangle = \frac{\Omega}{\tilde{\Omega} + \frac{1}{2}\Delta}\sqrt{\frac{\tilde{\Omega} + \frac{1}{2}\Delta}{2\tilde{\Omega}}}|1\rangle + \sqrt{\frac{\tilde{\Omega} + \frac{1}{2}\Delta}{2\tilde{\Omega}}}|2\rangle$$

$$= \sqrt{\frac{\Omega^2}{2\tilde{\Omega}(\tilde{\Omega} + \frac{1}{2}\Delta)}}|1\rangle + \sqrt{\frac{\tilde{\Omega} + \frac{1}{2}\Delta}{2\tilde{\Omega}}}|2\rangle$$

$$= \sqrt{\frac{\tilde{\Omega}^2 - \frac{1}{4}\Delta^2}{2\tilde{\Omega}(\tilde{\Omega} + \frac{1}{2}\Delta)}}|1\rangle + \sqrt{\frac{\tilde{\Omega} + \frac{1}{2}\Delta}{2\tilde{\Omega}}}|2\rangle$$

$$= \sqrt{\frac{\tilde{\Omega} - \frac{1}{2}\Delta}{2\tilde{\Omega}}}|1\rangle + \sqrt{\frac{\tilde{\Omega} + \frac{1}{2}\Delta}{2\tilde{\Omega}}}|2\rangle. \tag{12.34}$$

Similarly, for $\lambda_1 = -\tilde{\Omega}$, the eigenvalue equation is of the form

$$\begin{pmatrix} -\frac{1}{2}\Delta & \Omega \\ \Omega & \frac{1}{2}\Delta \end{pmatrix} \begin{pmatrix} c_1 \\ c_2 \end{pmatrix} = -\tilde{\Omega} \begin{pmatrix} c_1 \\ c_2 \end{pmatrix}, \qquad (12.35)$$

from which we find an equation relating the coefficients c_1 and c_2:

$$-\frac{1}{2}\Delta c_1 + \Omega c_2 = -\tilde{\Omega}c_1, \qquad (12.36)$$

From this relation, we find

$$c_1 = \frac{-\Omega}{\tilde{\Omega} - \frac{1}{2}\Delta} c_2. \qquad (12.37)$$

Thus, the energy state of the system corresponding to the energy $\lambda_1 = -\tilde{\Omega}$ is

$$|\Psi_2\rangle = \frac{-\Omega}{\tilde{\Omega} - \frac{1}{2}\Delta} c_2 |1\rangle + c_2 |2\rangle. \qquad (12.38)$$

From the normalization condition, $\langle \Psi_2 | \Psi_2 \rangle = 1$, we readily find

$$c_2 = \sqrt{\frac{\tilde{\Omega} - \frac{1}{2}\Delta}{2\tilde{\Omega}}}. \qquad (12.39)$$

Hence, the state $|\Psi_2\rangle$ takes the form

$$\begin{aligned}
|\Psi_2\rangle &= \frac{-\Omega}{\tilde{\Omega} - \frac{1}{2}\Delta} \sqrt{\frac{\tilde{\Omega} - \frac{1}{2}\Delta}{2\tilde{\Omega}}} |1\rangle + \sqrt{\frac{\tilde{\Omega} - \frac{1}{2}\Delta}{2\tilde{\Omega}}} |2\rangle \\
&= -\sqrt{\frac{\Omega^2}{2\tilde{\Omega}(\tilde{\Omega} - \frac{1}{2}\Delta)}} |1\rangle + \sqrt{\frac{\tilde{\Omega} - \frac{1}{2}\Delta}{2\tilde{\Omega}}} |2\rangle \\
&= -\sqrt{\frac{\tilde{\Omega}^2 - \frac{1}{4}\Delta^2}{2\tilde{\Omega}(\tilde{\Omega} - \frac{1}{2}\Delta)}} |1\rangle + \sqrt{\frac{\tilde{\Omega} - \frac{1}{2}\Delta}{2\tilde{\Omega}}} |2\rangle \\
&= -\sqrt{\frac{\tilde{\Omega} + \frac{1}{2}\Delta}{2\tilde{\Omega}}} |1\rangle + \sqrt{\frac{\tilde{\Omega} - \frac{1}{2}\Delta}{2\tilde{\Omega}}} |2\rangle. \qquad (12.40)
\end{aligned}$$

It is easy to check that the energy states $|\Psi_1\rangle$ and $|\Psi_2\rangle$ are orthogonal, i.e., $\langle \Psi_1 | \Psi_2 \rangle = 0$.

The energy states are linear superpositions of the atomic states. The superpositions are induced by the laser field forcing the electron to jump between the atomic states $|1\rangle$ and $|2\rangle$. It is often said that the superpositions result from "dressing" the atom in the laser field. For this reason, the states are called in the literature as dressed states of the two-level system.

Chapter 13

Spin Operators and Pauli Matrices

Problem 13.1

Calculate the square $(\vec{A} \cdot \vec{S})^2$ of the scalar product of an arbitrary vector $\vec{A}$ and of the spin vector $\vec{S} = S_x \vec{i} + S_y \vec{j} + S_z \vec{k}$.

Solution

In the cartesian coordinates, $\vec{A} = A_x \vec{i} + A_y \vec{j} + A_z \vec{k}$. Hence, the dot product between $\vec{A}$ and $\vec{S}$ is

$$\vec{A} \cdot \vec{S} = A_x S_x + A_y S_y + A_z S_z. \tag{13.1}$$

Squaring $\vec{A} \cdot \vec{S}$, we obtain

$$
\begin{aligned}
\left(\vec{A} \cdot \vec{S}\right)^2 &= \left(A_x S_x + A_y S_y + A_z S_z\right)\left(A_x S_x + A_y S_y + A_z S_z\right) \\
&= A_x^2 S_x^2 + A_x A_y S_x S_y + A_x A_z S_x S_z \\
&\quad + A_y A_x S_y S_x + A_y^2 S_y^2 + A_y A_z S_y S_z \\
&\quad + A_z A_x S_z S_x + A_z A_y S_z S_y + A_z^2 S_z^2 \\
&= A_x^2 S_x^2 + A_y^2 S_y^2 + A_z^2 S_z^2 + A_x A_y [S_x, S_y]_+ \\
&\quad + A_x A_z [S_x, S_z]_+ + A_y A_z [S_y, S_z]_+.
\end{aligned}
\tag{13.2}
$$

Problems and Solutions in Quantum Physics
Zbigniew Ficek
Copyright © 2016 Pan Stanford Publishing Pte. Ltd.
ISBN 978-981-4669-36-8 (Hardcover), 978-981-4669-37-5 (eBook)
www.panstanford.com

Since the components of the spin anticommute, i.e.,

$$[S_x, S_y]_+ = [S_x, S_z]_+ = [S_y, S_z]_+ = 0 \qquad (13.3)$$

and

$$S_x^2 = S_y^2 = S_z^2 = \frac{1}{4}\hbar^2, \qquad (13.4)$$

we finally get

$$\left(\vec{A} \cdot \vec{S}\right)^2 = \frac{1}{4}\hbar^2 \left(A_x^2 + A_y^2 + A_z^2\right) = \frac{1}{4}\hbar^2 |\vec{A}|^2 = |\vec{S}|^2 |\vec{A}|^2. \qquad (13.5)$$

Thus, $(\vec{A} \cdot \vec{S})^2$ is equal to the product of the squares of the vector $\vec{A}$ and the spin vector $\vec{S}$.

Problem 13.2

Calculate the squares of the spin components σ_x^2, σ_y^2, and σ_z^2, and verify if the squares of the spin components can be simultaneously measured with the same precision.

Solution

Calculate $\hat{\sigma}_x^2$:

$$\hat{\sigma}_x^2 = \hat{\sigma}_x \hat{\sigma}_x = \begin{pmatrix} 0 & 1 \\ 1 & 0 \end{pmatrix} \begin{pmatrix} 0 & 1 \\ 1 & 0 \end{pmatrix} = \begin{pmatrix} 1 & 0 \\ 0 & 1 \end{pmatrix} = \hat{1}. \qquad (13.6)$$

Similarly

$$\hat{\sigma}_y^2 = \hat{\sigma}_y \hat{\sigma}_y = \begin{pmatrix} 0 & -i \\ i & 0 \end{pmatrix} \begin{pmatrix} 0 & -i \\ i & 0 \end{pmatrix} = \begin{pmatrix} 1 & 0 \\ 0 & 1 \end{pmatrix} = \hat{1}, \qquad (13.7)$$

and

$$\hat{\sigma}_z^2 = \hat{\sigma}_z \hat{\sigma}_z = \begin{pmatrix} 1 & 0 \\ 0 & -1 \end{pmatrix} \begin{pmatrix} 1 & 0 \\ 0 & -1 \end{pmatrix} = \begin{pmatrix} 1 & 0 \\ 0 & 1 \end{pmatrix} = \hat{1}. \qquad (13.8)$$

The squares of the spin components are unit operators. Hence, they commute with each other. Therefore, the squares of the spin components can be simultaneously measured with the same precision.

Problem 13.3

Matrix representation of spin operators

The operators $\hat{\sigma}_x$, $\hat{\sigma}_y$, and $\hat{\sigma}_z$ representing the components of the electron spin, can Ψ be written in terms of the spin raising and spin lowering operators σ^+ and σ^- as

$$\hat{\sigma}_x = \hat{\sigma}^+ + \hat{\sigma}^-,$$
$$\hat{\sigma}_y = \left(\hat{\sigma}^+ - \hat{\sigma}^-\right)/i,$$
$$\hat{\sigma}_z = \hat{\sigma}^+\hat{\sigma}^- - \hat{\sigma}^-\hat{\sigma}^+. \tag{13.9}$$

Let $|1\rangle$ and $|2\rangle$ the two eigenstates of the electron spin with the eigenvalues $-\hbar/2$ and $+\hbar/2$, respectively, as determined in the Stern–Gerlach experiment. The raising and lowering operators satisfy the following relations:

$$\hat{\sigma}^+|1\rangle = |2\rangle, \qquad \hat{\sigma}^-|1\rangle = 0,$$
$$\hat{\sigma}^+|2\rangle = 0, \qquad \hat{\sigma}^-|2\rangle = |1\rangle. \tag{13.10}$$

Using these relations, find the matrix representations (the Pauli matrices) of the operators $\hat{\sigma}_x$, $\hat{\sigma}_y$, and $\hat{\sigma}_z$ in the basis of the states $|1\rangle$ and $|2\rangle$.

Solution

Find first the matrix representation of $\hat{\sigma}_x$. In the basis of the states $|2\rangle$ and $|1\rangle$, the operator $\hat{\sigma}_x$ has a matrix representation of the form

$$\hat{\sigma}_x = \begin{pmatrix} \langle 2|\hat{\sigma}_x|2\rangle & \langle 2|\hat{\sigma}_x|1\rangle \\ \langle 1|\hat{\sigma}_x|2\rangle & \langle 1|\hat{\sigma}_x|1\rangle \end{pmatrix}. \tag{13.11}$$

The matrix elements are

$$\langle 1|\hat{\sigma}_x|1\rangle = \langle 1|\left(\hat{\sigma}^+ + \hat{\sigma}^-\right)|1\rangle = \langle 1|2\rangle + \langle 1|0 = 0 + 0 = 0,$$
$$\langle 1|\hat{\sigma}_x|2\rangle = \langle 1|\left(\hat{\sigma}^+ + \hat{\sigma}^-\right)|2\rangle = \langle 1|0 + \langle 1|1\rangle = 0 + 1 = 1,$$
$$\langle 2|\hat{\sigma}_x|1\rangle = \langle 2|\left(\hat{\sigma}^+ + \hat{\sigma}^-\right)|1\rangle = \langle 2|2\rangle + \langle 2|0 = 1 + 0 = 1,$$
$$\langle 2|\hat{\sigma}_x|2\rangle = \langle 2|\left(\hat{\sigma}^+ + \hat{\sigma}^-\right)|2\rangle = \langle 2|0 + \langle 2|1\rangle = 0 + 0 = 0,$$
$$\tag{13.12}$$

and then the matrix (13.11) takes the form

$$\hat{\sigma}_x = \begin{pmatrix} 0 & 1 \\ 1 & 0 \end{pmatrix}. \tag{13.13}$$

Similarly, in the basis of the states $|2\rangle$ and $|1\rangle$, the operator $\hat{\sigma}_y$ has a matrix representation of the form

$$\hat{\sigma}_y = \begin{pmatrix} \langle 2|\hat{\sigma}_y|2\rangle & \langle 2|\hat{\sigma}_y|1\rangle \\ \langle 1|\hat{\sigma}_y|2\rangle & \langle 1|\hat{\sigma}_y|1\rangle \end{pmatrix}. \tag{13.14}$$

Since

$$\langle 1|\hat{\sigma}_y|1\rangle = \langle 1| \left(\hat{\sigma}^+ - \hat{\sigma}^- \right) /i|1\rangle = (\langle 1|2\rangle - \langle 1|0\rangle/i = 0,$$
$$\langle 1|\hat{\sigma}_y|2\rangle = \langle 1| \left(\hat{\sigma}^+ - \hat{\sigma}^- \right) /i|2\rangle = (\langle 1|0 - \langle 1|1\rangle)/i = i,$$
$$\langle 2|\hat{\sigma}_y|1\rangle = \langle 2| \left(\hat{\sigma}^+ - \hat{\sigma}^- \right) /i|1\rangle = (\langle 2|2\rangle - \langle 2|0\rangle/i = -i,$$
$$\langle 2|\hat{\sigma}_y|2\rangle = \langle 2| \left(\hat{\sigma}^+ - \hat{\sigma}^- \right) /i|2\rangle = (\langle 2|0 - \langle 2|1\rangle)/i = 0, \tag{13.15}$$

the matrix (13.14) takes the form

$$\hat{\sigma}_y = \begin{pmatrix} 0 & -i \\ i & 0 \end{pmatrix}. \tag{13.16}$$

Finally, in the basis of the states $|2\rangle$ and $|1\rangle$, the operator $\hat{\sigma}_z$ has a matrix representation of the form

$$\hat{\sigma}_z = \begin{pmatrix} \langle 2|\hat{\sigma}_z|2\rangle & \langle 2|\hat{\sigma}_z|1\rangle \\ \langle 1|\hat{\sigma}_z|2\rangle & \langle 1|\hat{\sigma}_z|1\rangle \end{pmatrix}. \tag{13.17}$$

Since

$$\langle 1|\hat{\sigma}_z|1\rangle = \langle 1| \left(\hat{\sigma}^+\hat{\sigma}^- - \hat{\sigma}^-\hat{\sigma}^+ \right) |1\rangle = \langle 1|0 - \langle 1|1\rangle = -1,$$
$$\langle 1|\hat{\sigma}_z|2\rangle = \langle 1| \left(\hat{\sigma}^+\hat{\sigma}^- - \hat{\sigma}^-\hat{\sigma}^+ \right) |2\rangle = \langle 1|2\rangle - \langle 1|0 = 0,$$
$$\langle 2|\hat{\sigma}_z|1\rangle = \langle 2| \left(\hat{\sigma}^+\hat{\sigma}^- - \hat{\sigma}^-\hat{\sigma}^+ \right) |1\rangle = \langle 2|0 - \langle 2||1\rangle = 0,$$
$$\langle 2|\hat{\sigma}_z|2\rangle = \langle 2| \left(\hat{\sigma}^+\hat{\sigma}^- - \hat{\sigma}^-\hat{\sigma}^+ \right) |2\rangle = \langle 2|2\rangle - \langle 2|0 = 1, \tag{13.18}$$

the matrix (13.17) takes the form

$$\hat{\sigma}_z = \begin{pmatrix} 1 & 0 \\ 0 & -1 \end{pmatrix}. \tag{13.19}$$

Problem 13.4

Properties of the Pauli matrices

Consider the Pauli matrices representing the spin operators $\hat{\sigma}_x$, $\hat{\sigma}_y$, and $\hat{\sigma}_z$ in the basis of the states $|1\rangle$ and $|2\rangle$.

(a) Prove that the operators $\hat{\sigma}_x$, $\hat{\sigma}_y$, $\hat{\sigma}_z$ are Hermitian. This result is what the student could expect as the operators represent a physical (measurable) quantity, the electron spin.

(b) Show that each of the operators $\hat{\sigma}_x$, $\hat{\sigma}_y$, $\hat{\sigma}_z$ has eigenvalues $+1$, -1. Determine the normalized eigenvectors of each. Are $|1\rangle$ and $|2\rangle$ the eigenvectors of any of the matrices?

(c) Show that the operators $\hat{\sigma}_x$, $\hat{\sigma}_y$, $\hat{\sigma}_z$ obey the commutation relations

$$\left[\hat{\sigma}_x, \hat{\sigma}_y\right] = 2i\hat{\sigma}_z,$$
$$\left[\hat{\sigma}_z, \hat{\sigma}_x\right] = 2i\hat{\sigma}_y,$$
$$\left[\hat{\sigma}_y, \hat{\sigma}_z\right] = 2i\hat{\sigma}_x. \tag{13.20}$$

If you recall the Heisenberg uncertainty relation, you will conclude immediately that these commutation relations show that the three components of the spin cannot be measured simultaneously with the same precision.

(d) Calculate anticommutators $\left[\hat{\sigma}_x, \hat{\sigma}_y\right]_+$, $\left[\hat{\sigma}_z, \hat{\sigma}_x\right]_+$, $\left[\hat{\sigma}_y, \hat{\sigma}_z\right]_+$.

(e) Show that $\hat{\sigma}_x^2 = \hat{\sigma}_y^2 = \hat{\sigma}_z^2 = \hat{1}$. This result is a confirmation of the conservation of the total spin of the system that the magnitude of the total spin vector is constant.

(f) Write the operators $\hat{\sigma}_x$, $\hat{\sigma}_y$, and $\hat{\sigma}_z$ in terms of the projection operators $\hat{P}_{ij} = |i\rangle\langle j|$, $(i, j = 1, 2)$.

Solution (a)

An operator (matrix) $\hat{A}$ is Hermitian if

$$\langle\phi_i|\hat{A}|\phi_j\rangle = \langle\phi_j|\hat{A}|\phi_i\rangle^*. \tag{13.21}$$

It is easy to see that for all the matrices

$$\langle\phi_i|\hat{\sigma}_n|\phi_j\rangle = \langle\phi_j|\hat{\sigma}_n|\phi_i\rangle^*, \qquad i, j = 1, 2, \tag{13.22}$$

where $n = x, y, z$. Thus, the matrices $\hat{\sigma}_x$, $\hat{\sigma}_y$, $\hat{\sigma}_z$ are Hermitian.

Solution (b)

Consider an eigenvalue equation for $\hat{\sigma}_x$:

$$\begin{pmatrix} 0 & 1 \\ 1 & 0 \end{pmatrix} \begin{pmatrix} c_1 \\ c_2 \end{pmatrix} = \lambda \begin{pmatrix} c_1 \\ c_2 \end{pmatrix}, \tag{13.23}$$

which can be written as

$$\begin{pmatrix} -\lambda & 1 \\ 1 & -\lambda \end{pmatrix} \begin{pmatrix} c_1 \\ c_2 \end{pmatrix} = 0. \tag{13.24}$$

This equation has nonzero solutions when the determinant of the matrix is zero, i.e., when

$$\begin{vmatrix} -\lambda & 1 \\ 1 & -\lambda \end{vmatrix} = 0. \tag{13.25}$$

From this we find a quadratic equation

$$\lambda^2 - 1 = 0, \tag{13.26}$$

whose solutions are

$$\lambda = \pm 1. \tag{13.27}$$

Thus, the eigenvalues of the matrix $\hat{\sigma}_x$ are $+1$ and -1.

Now, we will find eigenvectors corresponding to the eigenvalues $\lambda = \pm 1$.

For $\lambda = 1$, the eigenvalue equation is of the form

$$\begin{pmatrix} 0 & 1 \\ 1 & 0 \end{pmatrix} \begin{pmatrix} c_1 \\ c_2 \end{pmatrix} = \begin{pmatrix} c_1 \\ c_2 \end{pmatrix}, \tag{13.28}$$

from which we find

$$c_1 = c_2. \tag{13.29}$$

Thus, the eigenvector of the matrix $\hat{\sigma}_x$ corresponding to the eigenvalue $+1$ is of the form

$$|\Psi_x\rangle_{+1} = c_1 \left(|\phi_1\rangle + |\phi_2\rangle \right). \tag{13.30}$$

We find the coefficient c_1 from the normalization condition

$$\begin{aligned} 1 &= {}_{+1}\langle \Psi_x | \Psi_x \rangle_{+1} = \left(\langle \phi_1 | + \langle \phi_2 | \right) c_1^* c_1 \left(|\phi_1\rangle + |\phi_2\rangle \right) \\ &= |c_1|^2 \left(\langle \phi_1 | \phi_1 \rangle + \langle \phi_1 | \phi_2 \rangle + \langle \phi_2 | \phi_1 \rangle + \langle \phi_2 | \phi_2 \rangle \right) \\ &= |c_1|^2 \left(1 + 0 + 0 + 1 \right) = 2|c_1|^2. \end{aligned} \tag{13.31}$$

Hence

$$c_1 = \frac{1}{\sqrt{2}},$$ (13.32)

and then the normalized eigenvector of $\hat{\sigma}_x$ corresponding to the eigenvalue $+1$ is of the form

$$|\Psi_x\rangle_{+1} = \frac{1}{\sqrt{2}} \left(|\phi_1\rangle + |\phi_2\rangle \right).$$ (13.33)

Similarly, for $\lambda = -1$, the eigenvalue equation is of the form

$$\begin{pmatrix} 0 & 1 \\ 1 & 0 \end{pmatrix} \begin{pmatrix} c_1 \\ c_2 \end{pmatrix} = - \begin{pmatrix} c_1 \\ c_2 \end{pmatrix},$$ (13.34)

from which we find that

$$c_2 = -c_1.$$ (13.35)

Thus, the eigenvector of the matrix $\hat{\sigma}_x$ corresponding to the eigenvalue -1 is of the form

$$|\Psi_x\rangle_{-1} = c_1 \left(|\phi_1\rangle - |\phi_2\rangle \right).$$ (13.36)

As usual, we find the coefficient c_1 from the normalization condition

$$
\begin{aligned}
1 &= {}_{-1}\langle \Psi_x | \Psi_x \rangle_{-1} = \left(\langle \phi_1 | - \langle \phi_2 | \right) c_1^* c_1 \left(|\phi_1\rangle - |\phi_2\rangle \right) \\
&= |c_1|^2 \left(\langle \phi_1 | \phi_1 \rangle - \langle \phi_1 | \phi_2 \rangle - \langle \phi_2 | \phi_1 \rangle + \langle \phi_2 | \phi_2 \rangle \right) \\
&= |c_1|^2 \left(1 - 0 - 0 + 1 \right) = 2|c_1|^2.
\end{aligned}
$$ (13.37)

Hence

$$c_1 = \frac{1}{\sqrt{2}},$$ (13.38)

and then the normalized eigenvector of $\hat{\sigma}_x$ corresponding to the eigenvalue -1 is of the form

$$|\Psi_x\rangle_{-1} = \frac{1}{\sqrt{2}} \left(|\phi_1\rangle - |\phi_2\rangle \right).$$ (13.39)

In summary, the normalized eigenvectors of $\hat{\sigma}_x$ written in terms of the orthonormal vectors $|\phi_1\rangle$ and $|\phi_2\rangle$ are of the form

$$|\Psi_x\rangle_{+1} = \frac{1}{\sqrt{2}} \left(|\phi_1\rangle + |\phi_2\rangle \right),$$

$$|\Psi_x\rangle_{-1} = \frac{1}{\sqrt{2}} \left(|\phi_1\rangle - |\phi_2\rangle \right).$$ (13.40)

Consider now the matrix $\hat{\sigma}_y$.

The eigenvalue equation for $\hat{\sigma}_y$ is of the form

$$\begin{pmatrix} 0 & -i \\ i & 0 \end{pmatrix} \begin{pmatrix} c_1 \\ c_2 \end{pmatrix} = \lambda \begin{pmatrix} c_1 \\ c_2 \end{pmatrix}, \qquad (13.41)$$

which can be written as

$$\begin{pmatrix} -\lambda & -i \\ i & -\lambda \end{pmatrix} \begin{pmatrix} c_1 \\ c_2 \end{pmatrix} = 0. \qquad (13.42)$$

This equation has nonzero solutions when the determinant of the matrix is zero, i.e., when

$$\begin{vmatrix} -\lambda & -i \\ i & -\lambda \end{vmatrix} = 0. \qquad (13.43)$$

From this we find that the determinant is equal to a quadratic equation

$$\lambda^2 - 1 = 0, \qquad (13.44)$$

whose solutions are

$$\lambda = \pm 1. \qquad (13.45)$$

Thus, the eigenvalues of the matrix $\hat{\sigma}_y$ are $+1$ and -1.

Now, we will find eigenvectors corresponding to the eigenvalues $\lambda = \pm 1$.

For $\lambda = 1$, the eigenvalue equation is of the form

$$\begin{pmatrix} 0 & -i \\ i & 0 \end{pmatrix} \begin{pmatrix} c_1 \\ c_2 \end{pmatrix} = \begin{pmatrix} c_1 \\ c_2 \end{pmatrix}, \qquad (13.46)$$

from which we find

$$c_1 = -ic_2. \qquad (13.47)$$

Thus, the eigenvector of the matrix $\hat{\sigma}_y$ corresponding to the eigenvalue $+1$ is of the form

$$|\Psi_y\rangle_{+1} = c_1 \left(|\phi_1\rangle + i|\phi_2\rangle \right). \qquad (13.48)$$

As usual, we find the coefficient c_1 from the normalization condition

$$\begin{aligned}
1 &= {}_{+1}\langle \Psi_y | \Psi_y \rangle_{+1} = \left(\langle\phi_1| - i\langle\phi_2| \right) c_1^* c_1 \left(|\phi_1\rangle + i|\phi_2\rangle \right) \\
&= |c_1|^2 \left(\langle\phi_1|\phi_1\rangle + i\langle\phi_1|\phi_2\rangle - i\langle\phi_2|\phi_1\rangle + \langle\phi_2|\phi_2\rangle \right) \\
&= |c_1|^2 \left(1 + i0 - i0 + 1 \right) = 2|c_1|^2. \qquad (13.49)
\end{aligned}$$

Hence

$$c_1 = \frac{1}{\sqrt{2}},\qquad(13.50)$$

and then the normalized eigenvector of $\hat{\sigma}_y$ corresponding to the eigenvalue $+1$ is of the form

$$|\Psi_y\rangle_{+1} = \frac{1}{\sqrt{2}}\left(|\phi_1\rangle + i|\phi_2\rangle\right).\qquad(13.51)$$

Similarly, for $\lambda = -1$, the eigenvalue equation is of the form

$$\begin{pmatrix} 0 & -i \\ i & 0 \end{pmatrix}\begin{pmatrix} c_1 \\ c_2 \end{pmatrix} = -\begin{pmatrix} c_1 \\ c_2 \end{pmatrix},\qquad(13.52)$$

from which we find

$$c_2 = -i c_1.\qquad(13.53)$$

Thus, the eigenvector of the matrix $\hat{\sigma}_y$ corresponding to the eigenvalue -1 is of the form

$$|\Psi_y\rangle_{-1} = c_1\left(|\phi_1\rangle - i|\phi_2\rangle\right).\qquad(13.54)$$

As usual, we find the coefficient c_1 from the normalization condition

$$\begin{aligned} 1 &= {}_{-1}\langle\,\Psi_y|\Psi_y\rangle_{-1} = \left(\langle\phi_1| + i\langle\phi_2|\right)c_1^* c_1\left(|\phi_1\rangle - i|\phi_2\rangle\right) \\ &= |c_1|^2\left(\langle\phi_1|\phi_1\rangle - i\langle\phi_1|\phi_2\rangle + i\langle\phi_2|\phi_1\rangle + \langle\phi_2|\phi_2\rangle\right) \\ &= |c_1|^2\left(1 + i0 - i0 + 1\right) = 2|c_1|^2. \end{aligned}\qquad(13.55)$$

Hence

$$c_1 = \frac{1}{\sqrt{2}},\qquad(13.56)$$

and then the normalized eigenvector of $\hat{\sigma}_y$ corresponding to the eigenvalue -1 is of the form

$$|\Psi_y\rangle_{-1} = \frac{1}{\sqrt{2}}\left(|\phi_1\rangle - i|\phi_2\rangle\right).\qquad(13.57)$$

In summary, the normalized eigenvectors of $\hat{\sigma}_y$ written in terms of the orthonormal vectors $|\phi_1\rangle$ and $|\phi_2\rangle$ are of the form

$$|\Psi_y\rangle_{+1} = \frac{1}{\sqrt{2}}\left(|\phi_1\rangle + i|\phi_2\rangle\right),$$

$$|\Psi_y\rangle_{-1} = \frac{1}{\sqrt{2}}\left(|\phi_1\rangle - i|\phi_2\rangle\right).\qquad(13.58)$$

Finally, consider the matrix $\hat{\sigma}_z$.

It is easily to see that the matrix $\hat{\sigma}_z$ is diagonal. Thus, the basis vectors $|\phi_1\rangle$ and $|\phi_2\rangle$ are the eigenvectors of $\hat{\sigma}_z$. Since

$$\langle\phi_1|\hat{\sigma}_z|\phi_1\rangle = 1 \qquad \text{and} \qquad \langle\phi_2|\hat{\sigma}_z|\phi_2\rangle = -1,\qquad(13.59)$$

we see that $|\phi_1\rangle$ is an eigenvector of $\hat{\sigma}_z$ with eigenvalue $+1$, and $|\phi_2\rangle$ is an eigenvector of $\hat{\sigma}_z$ with eigenvalue -1.

Solution (c)

Consider the commutator $\left[\hat{\sigma}_x, \hat{\sigma}_y\right]$:

$$\left[\hat{\sigma}_x, \hat{\sigma}_y\right] = \hat{\sigma}_x\hat{\sigma}_y - \hat{\sigma}_y\hat{\sigma}_x$$

$$= \begin{pmatrix} 0 & 1 \\ 1 & 0 \end{pmatrix}\begin{pmatrix} 0 & -i \\ i & 0 \end{pmatrix} - \begin{pmatrix} 0 & -i \\ i & 0 \end{pmatrix}\begin{pmatrix} 0 & 1 \\ 1 & 0 \end{pmatrix}$$

$$= \begin{pmatrix} i & 0 \\ 0 & -i \end{pmatrix} - \begin{pmatrix} -i & 0 \\ 0 & i \end{pmatrix} = \begin{pmatrix} 2i & 0 \\ 0 & -2i \end{pmatrix}$$

$$= 2i\begin{pmatrix} 1 & 0 \\ 0 & -1 \end{pmatrix} = 2i\hat{\sigma}_z. \tag{13.60}$$

Similarly,

$$\left[\hat{\sigma}_z, \hat{\sigma}_x\right] = \hat{\sigma}_z\hat{\sigma}_x - \hat{\sigma}_x\hat{\sigma}_z$$

$$= \begin{pmatrix} 1 & 0 \\ 0 & -1 \end{pmatrix}\begin{pmatrix} 0 & 1 \\ 1 & 0 \end{pmatrix} - \begin{pmatrix} 0 & 1 \\ 1 & 0 \end{pmatrix}\begin{pmatrix} 1 & 0 \\ 0 & -1 \end{pmatrix}$$

$$= \begin{pmatrix} 0 & 1 \\ -1 & 0 \end{pmatrix} - \begin{pmatrix} 0 & -1 \\ 1 & 0 \end{pmatrix} = \begin{pmatrix} 0 & 2 \\ -2 & 0 \end{pmatrix}$$

$$= 2i\begin{pmatrix} 0 & -i \\ i & 0 \end{pmatrix} = 2i\hat{\sigma}_y, \tag{13.61}$$

and

$$\left[\hat{\sigma}_y, \hat{\sigma}_z\right] = \hat{\sigma}_y\hat{\sigma}_z - \hat{\sigma}_z\hat{\sigma}_y$$

$$= \begin{pmatrix} 0 & -i \\ i & 0 \end{pmatrix}\begin{pmatrix} 1 & 0 \\ 0 & -1 \end{pmatrix} - \begin{pmatrix} 1 & 0 \\ 0 & -1 \end{pmatrix}\begin{pmatrix} 0 & -i \\ i & 0 \end{pmatrix}$$

$$= \begin{pmatrix} 0 & i \\ i & 0 \end{pmatrix} - \begin{pmatrix} 0 & -i \\ -i & 0 \end{pmatrix} = \begin{pmatrix} 0 & 2i \\ 2i & 0 \end{pmatrix}$$

$$= 2i\begin{pmatrix} 0 & 1 \\ 1 & 0 \end{pmatrix} = 2i\hat{\sigma}_x. \tag{13.62}$$

Solution (d)

Consider the anticommutator $[\hat{\sigma}_x, \hat{\sigma}_y]_+$:

$$[\hat{\sigma}_x, \hat{\sigma}_y]_+ = \hat{\sigma}_x\hat{\sigma}_y + \hat{\sigma}_y\hat{\sigma}_x$$

$$= \begin{pmatrix} 0 & 1 \\ 1 & 0 \end{pmatrix} \begin{pmatrix} 0 & -i \\ i & 0 \end{pmatrix} + \begin{pmatrix} 0 & -i \\ i & 0 \end{pmatrix} \begin{pmatrix} 0 & 1 \\ 1 & 0 \end{pmatrix}$$

$$= \begin{pmatrix} i & 0 \\ 0 & -i \end{pmatrix} + \begin{pmatrix} -i & 0 \\ 0 & i \end{pmatrix} = \begin{pmatrix} 0 & 0 \\ 0 & 0 \end{pmatrix} = 0. \qquad (13.63)$$

Similarly,

$$[\hat{\sigma}_z, \hat{\sigma}_x]_+ = \hat{\sigma}_z\hat{\sigma}_x + \hat{\sigma}_x\hat{\sigma}_z$$

$$= \begin{pmatrix} 1 & 0 \\ 0 & -1 \end{pmatrix} \begin{pmatrix} 0 & 1 \\ 1 & 0 \end{pmatrix} + \begin{pmatrix} 0 & 1 \\ 1 & 0 \end{pmatrix} \begin{pmatrix} 1 & 0 \\ 0 & -1 \end{pmatrix}$$

$$= \begin{pmatrix} 0 & 1 \\ -1 & 0 \end{pmatrix} + \begin{pmatrix} 0 & -1 \\ 1 & 0 \end{pmatrix} = \begin{pmatrix} 0 & 0 \\ 0 & 0 \end{pmatrix} = 0, \qquad (13.64)$$

and

$$[\hat{\sigma}_y, \hat{\sigma}_z]_+ = \hat{\sigma}_y\hat{\sigma}_z + \hat{\sigma}_z\hat{\sigma}_y$$

$$= \begin{pmatrix} 0 & -i \\ i & 0 \end{pmatrix} \begin{pmatrix} 1 & 0 \\ 0 & -1 \end{pmatrix} + \begin{pmatrix} 1 & 0 \\ 0 & -1 \end{pmatrix} \begin{pmatrix} 0 & -i \\ i & 0 \end{pmatrix}$$

$$= \begin{pmatrix} 0 & i \\ i & 0 \end{pmatrix} + \begin{pmatrix} 0 & -i \\ -i & 0 \end{pmatrix} = \begin{pmatrix} 0 & 0 \\ 0 & 0 \end{pmatrix} = 0. \qquad (13.65)$$

Hence,

$$[\hat{\sigma}_x, \hat{\sigma}_y]_+ = [\hat{\sigma}_z, \hat{\sigma}_x]_+ = [\hat{\sigma}_y, \hat{\sigma}_z]_+ = 0. \qquad (13.66)$$

Solution (e)

Consider $\hat{\sigma}_x^2$:

$$\hat{\sigma}_x^2 = \hat{\sigma}_x\hat{\sigma}_x = \begin{pmatrix} 0 & 1 \\ 1 & 0 \end{pmatrix} \begin{pmatrix} 0 & 1 \\ 1 & 0 \end{pmatrix} = \begin{pmatrix} 1 & 0 \\ 0 & 1 \end{pmatrix} = \hat{1}. \qquad (13.67)$$

Similarly,

$$\hat{\sigma}_y^2 = \hat{\sigma}_y\hat{\sigma}_y = \begin{pmatrix} 0 & -i \\ i & 0 \end{pmatrix} \begin{pmatrix} 0 & -i \\ i & 0 \end{pmatrix} = \begin{pmatrix} 1 & 0 \\ 0 & 1 \end{pmatrix} = \hat{1}, \qquad (13.68)$$

and

$$\hat{\sigma}_z^2 = \hat{\sigma}_z\hat{\sigma}_z = \begin{pmatrix} 1 & 0 \\ 0 & -1 \end{pmatrix}\begin{pmatrix} 1 & 0 \\ 0 & -1 \end{pmatrix} = \begin{pmatrix} 1 & 0 \\ 0 & 1 \end{pmatrix} = \hat{1}. \quad (13.69)$$

Hence,

$$\hat{\sigma}_x^2 = \hat{\sigma}_y^2 = \hat{\sigma}_z^2 = \hat{1}. \quad (13.70)$$

Solution (f)

As we have shown in lecture, an arbitrary operator $\hat{A}$ can be written in terms of the projector operators as

$$\hat{A} = \sum_{n,m} A_{mn}|m\rangle\langle n| = \sum_{n,m} A_{mn}\hat{P}_{mn}. \quad (13.71)$$

Thus, in the basis of the two orthonormal states $|1\rangle, |2\rangle$, the operator $\hat{\sigma}_x$ can be written as

$$\hat{\sigma}_x = \sum_{n,m=1}^{2} \sigma_{mn}^x|m\rangle\langle n| = \sigma_{11}^x|1\rangle\langle 1| + \sigma_{12}^x|1\rangle\langle 2| + \sigma_{21}^x|2\rangle\langle 1| + \sigma_{22}^x|2\rangle\langle 2|.$$

$$(13.72)$$

Since

$$\sigma_{11}^x = \langle 1|\hat{\sigma}_x|1\rangle = 0, \qquad \sigma_{22}^x = \langle 2|\hat{\sigma}_x|2\rangle = 0$$
$$\sigma_{12}^x = \langle 1|\hat{\sigma}_x|2\rangle = 1, \qquad \sigma_{21}^x = \langle 2|\hat{\sigma}_x|1\rangle = 1, \quad (13.73)$$

we find

$$\hat{\sigma}_x = |1\rangle\langle 2| + |2\rangle\langle 1|. \quad (13.74)$$

Following the same procedure, we find that the operator $\hat{\sigma}_y$ can be written as

$$\hat{\sigma}_y = \sum_{n,m=1}^{2} \sigma_{mn}^y|m\rangle\langle n| = \sigma_{11}^y|1\rangle\langle 1| + \sigma_{12}^y|1\rangle\langle 2| + \sigma_{21}^y|2\rangle\langle 1| + \sigma_{22}^y|2\rangle\langle 2|.$$

$$(13.75)$$

Since

$$\sigma_{11}^y = \langle 1|\hat{\sigma}_y|1\rangle = 0, \qquad \sigma_{22}^y = \langle 2|\hat{\sigma}_y|2\rangle = 0$$
$$\sigma_{12}^y = \langle 1|\hat{\sigma}_y|2\rangle = -i, \qquad \sigma_{21}^y = \langle 2|\hat{\sigma}_y|1\rangle = i, \quad (13.76)$$

we find

$$\hat{\sigma}_y = -i \left(|1\rangle\langle 2| - |2\rangle\langle 1| \right). \tag{13.77}$$

Similarly, the operator $\hat{\sigma}_z$ can be written as

$$\hat{\sigma}_z = \sum_{n,m=1}^{2} \sigma_{mn}^z |m\rangle\langle n| = \sigma_{11}^z |1\rangle\langle 1| + \sigma_{12}^z |1\rangle\langle 2| + \sigma_{21}^z |2\rangle\langle 1| + \sigma_{22}^z |2\rangle\langle 2|.$$

$$\tag{13.78}$$

Since

$$\sigma_{11}^z = \langle 1|\hat{\sigma}_z|1\rangle = 1, \qquad \sigma_{22}^z = \langle 2|\hat{\sigma}_z|2\rangle = -1$$
$$\sigma_{12}^z = \langle 1|\hat{\sigma}_z|2\rangle = 0, \qquad \sigma_{21}^z = \langle 2|\hat{\sigma}_z|1\rangle = 0, \tag{13.79}$$

we find that

$$\hat{\sigma}_z = |1\rangle\langle 1| - |2\rangle\langle 2|. \tag{13.80}$$

Chapter 14

Quantum Dynamics and Pictures

Problem 14.1

Consider a two-level atom of energy states $|1\rangle$ and $|2\rangle$ driven by a laser field. The atom can be represented as a spin-$\frac{1}{2}$ particle and the laser field can be treated as a classical field. The Hamiltonian of the system is given by

$$\hat{H} = \frac{1}{2}\hbar\omega_0\hat{\sigma}_z - \frac{1}{2}i\hbar\Omega\left(\hat{\sigma}^+ e^{-i\omega_L t} - \hat{\sigma}^- e^{i\omega_L t}\right), \qquad (14.1)$$

where Ω is the Rabi frequency of the laser field, ω_0 is the atomic transition frequency, ω_L is the laser frequency, and $\hat{\sigma}_z$, $\hat{\sigma}^+$ and $\hat{\sigma}^-$ are the spin operators defined as

$$\hat{\sigma}_z = |2\rangle\langle 2| - |1\rangle\langle 1|, \quad \hat{\sigma}^+ = |2\rangle\langle 1|, \quad \hat{\sigma}^- = |1\rangle\langle 2|. \qquad (14.2)$$

(a) Calculate the equation of motion for $\hat{\sigma}^-$.
(b) The equation of motion derived in (a) contains a time-dependent coefficient. Find a unitary operator that transforms $\hat{\sigma}^-$ into $\hat{\bar{\sigma}}^-$ whose equation of motion is free from time-dependent coefficients.

Problems and Solutions in Quantum Physics
Zbigniew Ficek
Copyright © 2016 Pan Stanford Publishing Pte. Ltd.
ISBN 978-981-4669-36-8 (Hardcover), 978-981-4669-37-5 (eBook)
www.panstanford.com

Solution (a)

The equation of motion for an operator is found using the Heisenberg equation of motion. For the operator $\hat{\sigma}^-$, the equation of motion is given by

$$\frac{d}{dt}\hat{\sigma}^- = \frac{i}{\hbar}\left[\hat{H}, \hat{\sigma}^-\right]. \tag{14.3}$$

Evaluating the commutator $[\hat{H}, \hat{\sigma}^-]$, we get

$$\left[\hat{H}, \hat{\sigma}^-\right] = \frac{1}{2}\hbar\omega_0\left[\hat{\sigma}_z, \hat{\sigma}^-\right] - \frac{1}{2}i\hbar\Omega\left[\left(\hat{\sigma}^+ e^{-i\omega_L t} - \hat{\sigma}^- e^{i\omega_L t}\right), \hat{\sigma}^-\right]. \tag{14.4}$$

Since

$$\left[\hat{\sigma}_z, \hat{\sigma}^-\right] = -2\hat{\sigma}^-, \quad \left[\hat{\sigma}^+, \hat{\sigma}^-\right] = \hat{\sigma}_z, \quad \left[\hat{\sigma}^-, \hat{\sigma}^-\right] = 0, \tag{14.5}$$

we get

$$\left[\hat{H}, \hat{\sigma}^-\right] = -\hbar\omega_0\hat{\sigma}^- - \frac{1}{2}i\hbar\Omega\hat{\sigma}_z e^{-i\omega_L t}. \tag{14.6}$$

Hence, the equation of motion for the operator $\hat{\sigma}^-$ is of the form

$$\frac{d}{dt}\hat{\sigma}^- = -i\omega_0\hat{\sigma}^- + \frac{1}{2}\Omega e^{-i\omega_L t}\hat{\sigma}_z. \tag{14.7}$$

The equation of motion is a differential equation with a time-dependent coefficient $\Omega e^{-i\omega_L t}$. It makes the equation difficult to solve. It can be simplified to a differential equation with time-independent coefficients by making a unitary transformation of the operators.

Solution (b)

The time-dependent coefficient in Eq. (14.7) oscillates with frequency ω_L. Therefore, the unitary operator that transforms the equation to an equation with time-independent coefficients should involve the frequency ω_L. Moreover, it should involve an operator of the system whose commutator with $\hat{\sigma}^-$ is equal to $\hat{\sigma}^-$. A unitary operator that satisfies those requirements is of the form

$$\hat{U}(t) = e^{\frac{1}{2}i\omega_L\hat{\sigma}_z t}. \tag{14.8}$$

Introducing a new "transformed" operator $\hat{\tilde{\sigma}}^- = \hat{U}^\dagger(t)\hat{\sigma}^-\hat{U}(t)$, we find that the unitary operator $\hat{U}(t)$ transforms $\hat{\sigma}^-$ into

$$\hat{\tilde{\sigma}}^- = \hat{U}^\dagger(t)\hat{\sigma}^-\hat{U}(t) = \left(1 - \frac{1}{2}i\omega_L\hat{\sigma}_z t + \dots\right)\hat{\sigma}^-\left(1 + \frac{1}{2}i\omega_L\hat{\sigma}_z t + \dots\right)$$

$$= \left(\hat{\sigma}^- - \frac{1}{2}i\omega_L\hat{\sigma}_z\hat{\sigma}^- t + \dots\right)\left(1 + \frac{1}{2}i\omega_L\hat{\sigma}_z t + \dots\right)$$

$$= \hat{\sigma}^- + \frac{1}{2}i\omega_L t\hat{\sigma}^-\hat{\sigma}_z - \frac{1}{2}i\omega_L t\hat{\sigma}_z\hat{\sigma}^- + \dots$$

$$= \hat{\sigma}^- - \frac{1}{2}i\omega_L t\left[\hat{\sigma}_z, \hat{\sigma}^-\right] + \dots = \hat{\sigma}^- + i\omega_L t\hat{\sigma}^- + \dots$$

$$= \hat{\sigma}^-(1 + i\omega_L t + \dots) = \hat{\sigma}^- e^{i\omega_L t}. \tag{14.9}$$

Hence, the equation of motion for $\hat{\tilde{\sigma}}^-$ is of the form

$$\frac{d}{dt}\hat{\tilde{\sigma}}^- = \frac{d}{dt}\left(\hat{\sigma}^- e^{i\omega_L t}\right) = \left(\frac{d}{dt}\hat{\sigma}^-\right)e^{-i\omega_L t} + \hat{\sigma}^-\left(\frac{d}{dt}e^{i\omega_L t}\right)$$

$$= \left(-i\omega_0\hat{\sigma}^- + \frac{1}{2}\Omega\hat{\sigma}_z e^{-i\omega_L t}\right)e^{i\omega_L t} + i\omega_L\hat{\sigma}^- e^{i\omega_L t}$$

$$= -i(\omega_0 - \omega_L)\hat{\sigma}^- e^{i\omega_L t} + \frac{1}{2}\Omega\hat{\sigma}_z = -i\Delta\hat{\tilde{\sigma}}^- + \frac{1}{2}\Omega\hat{\sigma}_z, \tag{14.10}$$

where $\Delta = \omega_0 - \omega_L$. The equation of motion for the transformed operator is a differential equation with time-independent coefficients.

Problem 14.2

The Hamiltonian of the two-level atom interacting with a classical laser field can be written as

$$\hat{H} = \hat{H}_0 + \hat{V}(t), \tag{14.11}$$

where

$$\hat{H}_0 = \frac{1}{2}\hbar\omega_0\hat{\sigma}_z$$

$$\hat{V}(t) = -\frac{1}{2}i\hbar\Omega\left(\hat{\sigma}^+ e^{-i\omega_L t} - \hat{\sigma}^- e^{i\omega_L t}\right). \tag{14.12}$$

(a) Transform $\hat{V}(t)$ into the interaction picture to find $\hat{V}_I = \hat{U}_0^\dagger\hat{V}(t)\hat{U}_0$.
(b) Find the equation of motion for $\hat{\sigma}^-$ in the interaction picture, i.e., find the equation of motion for $\hat{\sigma}_I^-(t) = \hat{U}_I^\dagger\hat{\sigma}^-\hat{U}_I$.

Solution (a)

With the Hamiltonian (14.11), the unitary operator $\hat{U}_0(t, t_0)$, defined as

$$\hat{U}_0(t, t_0) = e^{-\frac{i}{\hbar}\hat{H}_0(t-t_0)}, \tag{14.13}$$

takes the form

$$\hat{U}_0(t) = e^{-\frac{1}{2}i\omega_0\hat{\sigma}_z t}, \tag{14.14}$$

where for simplicity, we have assumed that the initial time $t_0 = 0$. Hence,

$$\begin{aligned}
\hat{V}_I = \hat{U}_0^\dagger \hat{V}(t)\hat{U}_0 &= e^{\frac{1}{2}i\omega_0\hat{\sigma}_z t}\,\hat{V}(t)\,e^{-\frac{1}{2}i\omega_0\hat{\sigma}_z t} \\
&= \left(1 + \frac{1}{2}i\omega_0\hat{\sigma}_z t + \dots\right)\hat{V}(t)\left(1 - \frac{1}{2}i\omega_0\hat{\sigma}_z t + \dots\right) \\
&= \left(\hat{V}(t) + \frac{1}{2}i\omega_0 t\hat{\sigma}_z\hat{V}(t) + \dots\right)\left(1 - \frac{1}{2}i\omega_0\hat{\sigma}_z t + \dots\right) \\
&= \hat{V}(t) + \frac{1}{2}i\omega_0 t\hat{\sigma}_z\hat{V}(t) - \frac{1}{2}i\omega_0 t\hat{V}(t)\hat{\sigma}_z + \dots \\
&= \hat{V}(t) + \frac{1}{2}i\omega_0 t\left[\hat{\sigma}_z, \hat{V}(t)\right] + \dots
\end{aligned} \tag{14.15}$$

Calculate the commutator $[\hat{\sigma}_z, \hat{V}(t)]$:

$$\begin{aligned}
\left[\hat{\sigma}_z, \hat{V}(t)\right] &= -\frac{1}{2}i\hbar\Omega\left[\hat{\sigma}_z, \left(\hat{\sigma}^+ e^{-i\omega_L t} - \hat{\sigma}^- e^{i\omega_L t}\right)\right] \\
&= -\frac{1}{2}i\hbar\Omega\left[\hat{\sigma}_z, \hat{\sigma}^+\right]e^{-i\omega_L t} + \frac{1}{2}i\hbar\Omega\left[\hat{\sigma}_z, \hat{\sigma}^-\right]e^{i\omega_L t}.
\end{aligned} \tag{14.16}$$

Since

$$\left[\hat{\sigma}_z, \hat{\sigma}^+\right] = 2\hat{\sigma}^+, \quad \left[\hat{\sigma}_z, \hat{\sigma}^-\right] = -2\hat{\sigma}^-, \tag{14.17}$$

we get

$$\left[\hat{\sigma}_z, \hat{V}(t)\right] = -i\hbar\Omega\hat{\sigma}^+ e^{-i\omega_L t} - i\hbar\Omega\hat{\sigma}^- e^{i\omega_L t}, \tag{14.18}$$

and then

$$\hat{V}_I = \hat{V}(t) + \frac{1}{2}i\omega_0 t \left(-i\hbar\Omega\hat{\sigma}^+ e^{-i\omega_L t} - i\hbar\Omega\hat{\sigma}^- e^{i\omega_L t}\right) + \ldots$$

$$= -\frac{1}{2}i\hbar\Omega \left(\hat{\sigma}^+ e^{-i\omega_L t} - \hat{\sigma}^- e^{i\omega_L t}\right)$$

$$+i\omega_0 t \left(-\frac{1}{2}i\hbar\Omega\hat{\sigma}^+ e^{-i\omega_L t} - \frac{1}{2}i\hbar\Omega\hat{\sigma}^- e^{i\omega_L t}\right) + \ldots$$

$$= -\frac{1}{2}i\hbar\Omega\hat{\sigma}^+ e^{-i\omega_L t} (1 + i\omega_0 t + \ldots)$$

$$+\frac{1}{2}i\hbar\Omega\hat{\sigma}^- e^{i\omega_L t} (1 - i\omega_0 t + \ldots)$$

$$= -\frac{1}{2}i\hbar\Omega\hat{\sigma}^+ e^{-i\omega_L t} e^{i\omega_0 t} + \frac{1}{2}i\hbar\Omega\hat{\sigma}^- e^{i\omega_L t} e^{-i\omega_0 t}$$

$$= -\frac{1}{2}i\hbar\Omega \left(\hat{\sigma}^+ e^{i\Delta t} - \hat{\sigma}^- e^{-i\Delta t}\right). \tag{14.19}$$

Solution (b)

The unitary operator $\hat{U}_I(t, t_0)$ involves a time-independent Hamiltonian $\hat{V}$. Therefore, we first transform $\hat{V}(t)$ into a time-independent form. Referring to part (a) of this tutorial problem, one can readily notice that the transformation could be done with a unitary operator of the form

$$\hat{U}_0(t) = e^{-\frac{1}{2}i\omega_L\hat{\sigma}_z t}, \tag{14.20}$$

which is of the form of the unitary operator (14.14) with ω_0 replaced by ω_L. Then, following the same way as in part (a) of the problem, one can easily show that

$$\hat{V} = \hat{U}_0^\dagger(t)\hat{V}(t)\hat{U}_0(t) = -\frac{1}{2}i\hbar\Omega\left(\hat{\sigma}^+ - \hat{\sigma}^-\right). \tag{14.21}$$

We can now define the unitary operator in the interaction picture

$$\hat{U}_I(t) = e^{\frac{i}{\hbar}\hat{V}t} = e^{\frac{1}{2}\Omega(\hat{\sigma}^+ - \hat{\sigma}^-)t} = e^{\frac{1}{2}i\Omega\hat{\sigma}_y t}, \tag{14.22}$$

where $\hat{\sigma}_y = (\hat{\sigma}^+ - \hat{\sigma}^-)/i$.

The equation of motion for $\hat{\sigma}^-$ is

$$\frac{d}{dt}\hat{\sigma}^- = -i\Delta\hat{\sigma}^- + \frac{1}{2}\Omega\hat{\sigma}_z. \tag{14.23}$$

Hence

$$\frac{d}{dt}\hat{\sigma}_I^-(t) = \frac{d}{dt}\left(\hat{U}_I^\dagger \hat{\sigma}^- \hat{U}_I\right)$$

$$= \left(\frac{d}{dt}\hat{U}_I^\dagger\right)\hat{\sigma}^- \hat{U}_I + \hat{U}_I^\dagger \left(\frac{d}{dt}\hat{\sigma}^-\right)\hat{U}_I + \hat{U}_I^\dagger \hat{\sigma}^- \left(\frac{d}{dt}\hat{U}_I\right).$$

(14.24)

Since

$$\frac{d}{dt}\hat{U}_I^\dagger = -\frac{1}{2}i\Omega\hat{\sigma}_y \hat{U}_I^\dagger \quad \text{and} \quad \frac{d}{dt}\hat{U}_I = \frac{1}{2}i\Omega\hat{\sigma}_y \hat{U}_I, \quad (14.25)$$

we get

$$\frac{d}{dt}\hat{\sigma}_I^-(t) = -\frac{1}{2}i\Omega\hat{U}_I^\dagger\left[\hat{\sigma}_y, \hat{\sigma}^-\right]\hat{U}_I + \hat{U}_I^\dagger\left(-i\Delta\hat{\sigma}^- + \frac{1}{2}\Omega\hat{\sigma}_z\right)\hat{U}_I$$

$$= -\frac{1}{2}\Omega\hat{U}_I^\dagger\hat{\sigma}_z\hat{U}_I - i\Delta\hat{\sigma}_I^- + \frac{1}{2}\Omega\hat{U}_I^\dagger\hat{\sigma}_z\hat{U}_I = -i\Delta\hat{\sigma}_I^-.$$

(14.26)

Chapter 15

Quantum Harmonic Oscillator

Problem 15.1

Use the operator approach developed in Chapter 15 of textbook to prove that the nth harmonic oscillator energy eigenfunction obeys the following uncertainty relation

$$\delta x \delta p = \frac{\hbar}{2}(2n+1), \qquad (15.1)$$

where $\delta x = \sqrt{\langle \hat{x}^2 \rangle - \langle \hat{x} \rangle^2}$ and $\delta p_x = \sqrt{\langle \hat{p}_x^2 \rangle - \langle \hat{p}_x \rangle^2}$ are fluctuations of the position and momentum operators, respectively.

Solution

From the description of the position and momentum operators in terms of the annihilation and creation operators

$$\hat{x} = \frac{1}{2}\sqrt{\frac{2\hbar}{m\omega}}\left(\hat{a} + \hat{a}^\dagger\right), \quad \hat{p} = -i\sqrt{\frac{m\omega\hbar}{2}}\left(\hat{a} - \hat{a}^\dagger\right), \quad (15.2)$$

we have for the average value of the position operator in the nth energy state

$$\langle \hat{x} \rangle = \langle \phi_n | \hat{x} | \phi_n \rangle = A\left(\langle \phi_n | \hat{a} | \phi_n \rangle + \langle \phi_n | \hat{a}^\dagger | \phi_n \rangle\right), \quad (15.3)$$

Problems and Solutions in Quantum Physics
Zbigniew Ficek
Copyright © 2016 Pan Stanford Publishing Pte. Ltd.
ISBN 978-981-4669-36-8 (Hardcover), 978-981-4669-37-5 (eBook)
www.panstanford.com

where for simplicity, we have introduced a notation

$$A = \frac{1}{2}\sqrt{\frac{2\hbar}{m\omega}}. \tag{15.4}$$

However,

$$\hat{a}|\phi_n\rangle = \sqrt{n}|\phi_{n-1}\rangle, \quad \hat{a}^\dagger|\phi_n\rangle = \sqrt{n+1}|\phi_{n+1}\rangle, \tag{15.5}$$

and since $\langle\phi_n|\phi_{n\pm1}\rangle = 0$, we obtain

$$\langle\hat{x}\rangle = A\left(\sqrt{n}\langle\phi_n|\phi_{n-1}\rangle + \sqrt{n+1}\langle\phi_n|\phi_{n+1}\rangle\right) = 0. \tag{15.6}$$

In the same way, it is easily shown that

$$\langle\hat{p}\rangle = B\left(\sqrt{n}\langle\phi_n|\phi_{n-1}\rangle - \sqrt{n+1}\langle\phi_n|\phi_{n+1}\rangle\right) = 0, \tag{15.7}$$

where

$$B = -i\sqrt{\frac{m\omega\hbar}{2}}. \tag{15.8}$$

Next, we calculate $\langle\hat{x}^2\rangle$:

$$\begin{aligned}\langle\hat{x}^2\rangle &= \langle\phi_n|\hat{x}^2|\phi_n\rangle = A^2\langle\phi_n|\left(\hat{a}+\hat{a}^\dagger\right)\left(\hat{a}+\hat{a}^\dagger\right)|\phi_n\rangle \\ &= A^2\langle\phi_n|\hat{a}\hat{a} + \hat{a}^\dagger\hat{a} + \hat{a}\hat{a}^\dagger + \hat{a}^\dagger\hat{a}^\dagger|\phi_n\rangle.\end{aligned} \tag{15.9}$$

However,

$$\begin{aligned}\hat{a}\hat{a}|\phi_n\rangle &= \sqrt{n(n-1)}|\phi_{n-2}\rangle, \\ \hat{a}^\dagger\hat{a}|\phi_n\rangle &= n|\phi_n\rangle, \\ \hat{a}\hat{a}^\dagger|\phi_n\rangle &= (n+1)|\phi_n\rangle, \\ \hat{a}^\dagger\hat{a}^\dagger|\phi_n\rangle &= \sqrt{(n+1)(n+2)}|\phi_{n+2}\rangle.\end{aligned} \tag{15.10}$$

Thus, we obtain

$$\begin{aligned}\langle\hat{x}^2\rangle = A^2\left(\sqrt{n(n-1)}\langle\phi_n|\phi_{n-2}\rangle + n\langle\phi_n|\phi_n\rangle + (n+1)\langle\phi_n|\phi_n\rangle \right. \\ \left. + \sqrt{(n+1)(n+2)}\langle\phi_n|\phi_{n+2}\rangle\right).\end{aligned} \tag{15.11}$$

Since $\langle\phi_n|\phi_n\rangle = 1$ and $\langle\phi_n|\phi_{n\pm2}\rangle = 0$, we finally obtain

$$\langle\hat{x}^2\rangle = A^2(2n+1) = \frac{1}{2}\frac{\hbar}{m\omega}(2n+1). \tag{15.12}$$

Similarly for $\langle \hat{p}^2 \rangle$

$$\langle \hat{p}^2 \rangle = \langle \phi_n | \hat{p}^2 | \phi_n \rangle = B^2 \langle \phi_n | (\hat{a} - \hat{a}^\dagger)(\hat{a} - \hat{a}^\dagger) | \phi_n \rangle$$

$$= B^2 \langle \phi_n | \hat{a}\hat{a} - \hat{a}^\dagger \hat{a} - \hat{a}\hat{a}^\dagger + \hat{a}^\dagger \hat{a}^\dagger | \phi_n \rangle$$

$$= B^2 \left(\sqrt{n(n-1)} \langle \phi_n | \phi_{n-2} \rangle - n \langle \phi_n | \phi_n \rangle - (n+1) \langle \phi_n | \phi_n \rangle \right.$$

$$\left. + \sqrt{(n+1)(n+2)} \langle \phi_n | \phi_{n+2} \rangle \right) = -B^2(2n+1)$$

$$= - \left(-i\sqrt{\frac{m\omega\hbar}{2}} \right)^2 (2n+1) = \frac{m\omega\hbar}{2}(2n+1). \qquad (15.13)$$

Hence

$$\delta x \delta p = \sqrt{\frac{1}{2}\frac{\hbar}{m\omega}} \sqrt{\frac{m\omega\hbar}{2}}(2n+1) = \sqrt{\frac{\hbar^2}{4}}(2n+1) = \frac{\hbar}{2}(2n+1).$$

$$(15.14)$$

Problem 15.2

Given that $\hat{a}|n\rangle = \sqrt{n}|n-1\rangle$, show that n must be a positive integer.

Solution

Let $\hat{a}|n\rangle \equiv |\Psi\rangle$. Since the scalar product $\langle \Psi | \Psi \rangle \geq 0$, we have

$$\langle \Psi | \Psi \rangle = \langle n-1| \hat{a}^\dagger \hat{a} |n-1\rangle = n\langle n-1|n-1\rangle = n \geq 0. \quad (15.15)$$

Clearly, n is a positive integer.

Problem 15.3

(a) Using the commutation relation for the position $\hat{x}$ and momentum $\hat{p} \equiv \hat{p}_x$ operators

$$[\hat{x}, \hat{p}] = i\hbar, \qquad (15.16)$$

show that the annihilation and creation operators $\hat{a}$ and $\hat{a}^\dagger$ of a one-dimensional Harmonic oscillator satisfy the commutation relation

$$[\hat{a}, \hat{a}^\dagger] = \hat{1}. \qquad (15.17)$$

(b) Show that the Hamiltonian of the harmonic oscillator

$$\hat{H} = \frac{1}{2m}\hat{p}^2 + \frac{1}{2}m\omega^2\hat{x}^2 \qquad (15.18)$$

can be written as

$$\hat{H} = \hbar\omega\left(\hat{a}^\dagger\hat{a} + \frac{1}{2}\right). \qquad (15.19)$$

(c) Calculate the value of the uncertainty product $\Delta x\Delta p$ for a one-dimensional harmonic oscillator in the ground state $|\phi_0\rangle$, where $\Delta x = \sqrt{\langle\hat{x}^2\rangle - \langle\hat{x}\rangle^2}$ and $\Delta p = \sqrt{\langle\hat{p}^2\rangle - \langle\hat{p}\rangle^2}$.

Solution (a)

Since

$$\hat{a} = \sqrt{\frac{m\omega}{2\hbar}}\,\hat{x} + i\frac{1}{\sqrt{2m\hbar\omega}}\,\hat{p} = \alpha\hat{x} + i\beta\hat{p}, \qquad (15.20)$$

and the adjoint of this operator

$$\hat{a}^\dagger = \sqrt{\frac{m\omega}{2\hbar}}\,\hat{x} - i\frac{1}{\sqrt{2m\hbar\omega}}\,\hat{p} = \alpha\hat{x} - i\beta\hat{p}, \qquad (15.21)$$

where

$$\alpha = \sqrt{\frac{m\omega}{2\hbar}}, \qquad \beta = \frac{1}{\sqrt{2m\hbar\omega}}, \qquad (15.22)$$

we have

$$
\begin{aligned}
\left[\hat{a}, \hat{a}^\dagger\right] &= \hat{a}\hat{a}^\dagger - \hat{a}^\dagger\hat{a} = (\alpha\hat{x} + i\beta\hat{p})(\alpha\hat{x} - i\beta\hat{p}) \\
&\quad - (\alpha\hat{x} - i\beta\hat{p})(\alpha\hat{x} + i\beta\hat{p}) \\
&= \alpha^2\hat{x}^2 - i\alpha\beta\hat{x}\hat{p} + i\alpha\beta\hat{p}\hat{x} + \beta^2\hat{p}^2 - \alpha^2\hat{x}^2 \\
&\quad -i\alpha\beta\hat{x}\hat{p} + i\alpha\beta\hat{p}\hat{x} - \beta^2\hat{p}^2 \\
&= -2i\alpha\beta\hat{x}\hat{p} + 2i\alpha\beta\hat{p}\hat{x} = -2i\alpha\beta(\hat{x}\hat{p} - \hat{p}\hat{x}) = -2i\alpha\beta[\hat{x}, \hat{p}] \\
&= -2i\alpha\beta(i\hbar) = 2\alpha\beta\hbar = 2\hbar\sqrt{\frac{m\omega}{2\hbar}}\frac{1}{\sqrt{2m\hbar\omega}} = 1. \qquad (15.23)
\end{aligned}
$$

Solution (b)

Since

$$\hat{x} = \frac{1}{2}\sqrt{\frac{2\hbar}{m\omega}}\left(\hat{a} + \hat{a}^\dagger\right),$$

$$\hat{p} = -i\sqrt{\frac{m\omega\hbar}{2}}\left(\hat{a} - \hat{a}^\dagger\right), \tag{15.24}$$

and

$$\hat{H} = \frac{1}{2m}\hat{p}^2 + \frac{1}{2}m\omega^2\hat{x}^2, \tag{15.25}$$

we find

$$\begin{aligned}
\hat{H} &= -\frac{1}{2m}\frac{m\omega\hbar}{2}\left(\hat{a} - \hat{a}^\dagger\right)^2 + \frac{1}{2}m\omega^2\frac{1}{4}\frac{2\hbar}{m\omega}\left(\hat{a} + \hat{a}^\dagger\right)^2 \\
&= -\frac{1}{4}\hbar\omega\left[\left(\hat{a} - \hat{a}^\dagger\right)^2 - \left(\hat{a} + \hat{a}^\dagger\right)^2\right] \\
&= -\frac{1}{4}\hbar\omega\left(\hat{a}^2 - \hat{a}\hat{a}^\dagger - \hat{a}^\dagger\hat{a} + \hat{a}^{\dagger 2} - \hat{a}^2 - \hat{a}\hat{a}^\dagger - \hat{a}^\dagger\hat{a} - \hat{a}^{\dagger 2}\right) \\
&= \frac{1}{2}\hbar\omega\left(\hat{a}\hat{a}^\dagger + \hat{a}^\dagger\hat{a}\right). \tag{15.26}
\end{aligned}$$

From the commutation relation $\left[\hat{a}, \hat{a}^\dagger\right] = \hat{1}$, we get

$$\hat{H} = \frac{1}{2}\hbar\omega\left(\hat{a}\hat{a}^\dagger + \hat{a}^\dagger\hat{a}\right) = \frac{1}{2}\hbar\omega\left(\hat{a}^\dagger\hat{a} + 1 + \hat{a}^\dagger\hat{a}\right) = \hbar\omega\left(\hat{a}^\dagger\hat{a} + \frac{1}{2}\right). \tag{15.27}$$

Solution (c)

Since

$$\hat{x} = \frac{1}{2}\sqrt{\frac{2\hbar}{m\omega}}\left(\hat{a} + \hat{a}^\dagger\right),$$

$$\hat{p} = -i\sqrt{\frac{m\omega\hbar}{2}}\left(\hat{a} - \hat{a}^\dagger\right), \tag{15.28}$$

we have

$$\langle\hat{x}\rangle = \langle\phi_0|\hat{x}|\phi_0\rangle = \frac{1}{2}\sqrt{\frac{2\hbar}{m\omega}}\left(\langle\phi_0|\hat{a}|\phi_0\rangle + \langle\phi_0|\hat{a}^\dagger|\phi_0\rangle\right). \tag{15.29}$$

However,

$$\hat{a}|\phi_0\rangle = 0,$$
$$\hat{a}^\dagger|\phi_0\rangle = |\phi_1\rangle, \tag{15.30}$$

and since $\langle\phi_0|\phi_1\rangle = 0$, we obtain

$$\langle\hat{x}\rangle = \frac{1}{2}\sqrt{\frac{2\hbar}{m\omega}}\left(\langle\phi_0|0 + \langle\phi_0|\phi_1\rangle\right) = 0. \tag{15.31}$$

Similarly

$$\langle\hat{p}\rangle = 0. \tag{15.32}$$

Next, we calculate $\langle\hat{x}^2\rangle$ and $\langle\hat{p}^2\rangle$:

$$\langle\hat{x}^2\rangle = \langle\phi_0|\hat{x}^2|\phi_0\rangle = A^2\langle\phi_0|\left(\hat{a}+\hat{a}^\dagger\right)\left(\hat{a}+\hat{a}^\dagger\right)|\phi_0\rangle$$
$$= A^2\langle\phi_0|\hat{a}\hat{a} + \hat{a}^\dagger\hat{a} + \hat{a}\hat{a}^\dagger + \hat{a}^\dagger\hat{a}^\dagger|\phi_0\rangle, \tag{15.33}$$

where

$$A = \frac{1}{2}\sqrt{\frac{2\hbar}{m\omega}}. \tag{15.34}$$

However,

$$\hat{a}\hat{a}|\phi_0\rangle = 0,$$
$$\hat{a}^\dagger\hat{a}|\phi_0\rangle = 0,$$
$$\hat{a}\hat{a}^\dagger|\phi_0\rangle = |\phi_0\rangle,$$
$$\hat{a}^\dagger\hat{a}^\dagger|\phi_0\rangle = \sqrt{2}|\phi_2\rangle. \tag{15.35}$$

Thus, we obtain

$$\langle\hat{x}^2\rangle = A^2\left(0 + 0 + \langle\phi_0|\phi_0\rangle + \sqrt{2}\langle\phi_0|\phi_2\rangle\right). \tag{15.36}$$

Since $\langle\phi_0|\phi_0\rangle = 1$ and $\langle\phi_0|\phi_2\rangle = 0$, we finally obtain

$$\langle\hat{x}^2\rangle = A^2 = \frac{1}{2}\frac{\hbar}{m\omega}. \tag{15.37}$$

Similarly,

$$\langle\hat{p}^2\rangle = \langle\phi_0|\hat{p}^2|\phi_0\rangle = B^2\langle\phi_0|\left(\hat{a}-\hat{a}^\dagger\right)\left(\hat{a}-\hat{a}^\dagger\right)|\phi_0\rangle$$
$$= B^2\langle\phi_0|\hat{a}\hat{a} - \hat{a}^\dagger\hat{a} - \hat{a}\hat{a}^\dagger + \hat{a}^\dagger\hat{a}^\dagger|\phi_0\rangle$$
$$= B^2(0 - 0 - 1 + 0) = -\left(-i\sqrt{\frac{m\omega\hbar}{2}}\right)^2 = \frac{m\omega\hbar}{2}. \tag{15.38}$$

Hence,

$$\Delta x \Delta p = \sqrt{\frac{1}{2}\frac{\hbar}{m\omega}}\sqrt{\frac{m\omega\hbar}{2}} = \sqrt{\frac{\hbar^2}{4}} = \frac{\hbar}{2}. \tag{15.39}$$

Problem 15.4

Prove, by induction, the following commutation relation:

$$\left[\hat{a}, \left(\hat{a}^\dagger\right)^n\right] = n\left(\hat{a}^\dagger\right)^{n-1}. \qquad (15.40)$$

Solution

For $n = 1$,

$$\left[\hat{a}, \hat{a}^\dagger\right] = \hat{1}. \qquad (15.41)$$

Assume that the commutator is true for $n = k$:

$$\left[\hat{a}, \left(\hat{a}^\dagger\right)^k\right] = k\left(\hat{a}^\dagger\right)^{k-1}. \qquad (15.42)$$

We will show that the commutator is true for $n = k + 1$, i.e.,

$$\left[\hat{a}, \left(\hat{a}^\dagger\right)^{k+1}\right] = (k + 1)\left(\hat{a}^\dagger\right)^k. \qquad (15.43)$$

Consider the left-hand side of the above equation:

$$L = \left[\hat{a}, \left(\hat{a}^\dagger\right)^{k+1}\right] = \hat{a}\left(\hat{a}^\dagger\right)^{k+1} - \left(\hat{a}^\dagger\right)^{k+1}\hat{a} = \hat{a}\left(\hat{a}^\dagger\right)^k\hat{a}^\dagger - \left(\hat{a}^\dagger\right)^{k+1}\hat{a}$$

$$= \left(\hat{a}^\dagger\right)^k\hat{a}\hat{a}^\dagger + k\left(\hat{a}^\dagger\right)^{k-1}\hat{a}^\dagger - \left(\hat{a}^\dagger\right)^{k+1}\hat{a}$$

$$= \left(\hat{a}^\dagger\right)^k\left(1 + \hat{a}^\dagger\hat{a}\right) + k\left(\hat{a}^\dagger\right)^{k-1}\hat{a}^\dagger - \left(\hat{a}^\dagger\right)^{k+1}\hat{a}$$

$$= \left(\hat{a}^\dagger\right)^k + k\left(\hat{a}^\dagger\right)^k = (k + 1)\left(\hat{a}^\dagger\right)^k = R. \qquad (15.44)$$

Problem 15.5

Generation of an nth wave function from the ground state wave function

Using the normalized energy eigenfunctions of the Harmonic oscillator

$$|\phi_n\rangle = \frac{1}{\sqrt{n!}}\left(\hat{a}^\dagger\right)^n|\phi_0\rangle, \qquad (15.45)$$

show that

$$\hat{a}^\dagger|\phi_n\rangle = \sqrt{n + 1}|\phi_{n+1}\rangle,$$
$$\hat{a}|\phi_n\rangle = \sqrt{n}|\phi_{n-1}\rangle. \qquad (15.46)$$

Solution

Since

$$|\phi_1\rangle = \hat{a}^\dagger |\phi_0\rangle, \qquad (15.47)$$

we have

$$|\phi_2\rangle = \frac{1}{\sqrt{2}} \left(\hat{a}^\dagger\right)^2 |\phi_0\rangle = \frac{1}{\sqrt{2}} \hat{a}^\dagger |\phi_1\rangle. \qquad (15.48)$$

Hence,

$$\hat{a}^\dagger |\phi_1\rangle = \sqrt{2} |\phi_2\rangle. \qquad (15.49)$$

Next

$$|\phi_3\rangle = \frac{1}{\sqrt{3!}} \left(\hat{a}^\dagger\right)^3 |\phi_0\rangle = \frac{\sqrt{2}}{\sqrt{3!}} \hat{a}^\dagger |\phi_2\rangle = \frac{1}{\sqrt{3}} \hat{a}^\dagger |\phi_2\rangle. \quad (15.50)$$

Thus,

$$\hat{a}^\dagger |\phi_2\rangle = \sqrt{3} |\phi_3\rangle. \qquad (15.51)$$

Hence, we see from above that for an arbitrary n,

$$\hat{a}^\dagger |\phi_n\rangle = \sqrt{n+1} |\phi_{n+1}\rangle. \qquad (15.52)$$

Consider now the action of the annihilation operator on the wave function $|\phi_n\rangle$. Since

$$\hat{a}|\phi_0\rangle = 0, \qquad (15.53)$$

we get

$$\hat{a} |\phi_1\rangle = \hat{a}\hat{a}^\dagger |\phi_0\rangle = \left(1 + \hat{a}^\dagger \hat{a}\right) |\phi_0\rangle = |\phi_0\rangle. \qquad (15.54)$$

Thus,

$$\hat{a} |\phi_1\rangle = |\phi_0\rangle. \qquad (15.55)$$

Similarly,

$$\hat{a} |\phi_2\rangle = \frac{1}{\sqrt{2}} \hat{a}\hat{a}^\dagger |\phi_1\rangle = \frac{1}{\sqrt{2}} \left(1 + \hat{a}^\dagger \hat{a}\right) |\phi_1\rangle$$

$$= \frac{1}{\sqrt{2}} \left(|\phi_1\rangle + \hat{a}^\dagger |\phi_0\rangle\right) = \frac{1}{\sqrt{2}} \left(|\phi_1\rangle + |\phi_1\rangle\right) = \sqrt{2} |\phi_1\rangle.$$

$$(15.56)$$

Thus,

$$\hat{a}\,|\phi_2\rangle = \sqrt{2}\,|\phi_1\rangle,\qquad\qquad(15.57)$$

and in general

$$\hat{a}\,|\phi_n\rangle = \sqrt{n}\,|\phi_{n-1}\rangle.\qquad\qquad(15.58)$$

Problem 15.6

Matrix representation of the annihilation and creation operators

Write the matrix representations of the operators $\hat{a}$ and $\hat{a}^\dagger$ in the basis of the energy eigenstates $|\phi_n\rangle$, and using this representation, verify the commutation relation $[\hat{a}, \hat{a}^\dagger] = \hat{1}$, where $\hat{1}$ is the unit matrix.

Solution

Using the results of the Tutorial Problem 15.3, we can write the operators $\hat{a}$ and $\hat{a}^\dagger$ in the basis of the energy eigenstates $|\phi_n\rangle$ as

$$\hat{a} = \begin{pmatrix} 0 & \sqrt{1} & 0 & . & . & . \\ 0 & 0 & \sqrt{2} & 0 & . & . \\ 0 & 0 & 0 & \sqrt{3} & 0 & . \\ . & . & . & . & . & . \\ . & . & . & . & . & . \\ . & . & . & . & . & . \end{pmatrix},\qquad(15.59)$$

and similarly

$$\hat{a}^\dagger = \begin{pmatrix} 0 & . & . & . & . & . \\ \sqrt{1} & 0 & . & . & . & . \\ 0 & \sqrt{2} & 0 & . & . & . \\ 0 & 0 & \sqrt{3} & 0 & . & . \\ . & . & . & . & . & . \\ . & . & . & . & . & . \end{pmatrix}.\qquad(15.60)$$

Hence

$$
\hat{a}\hat{a}^{\dagger} =
\begin{pmatrix}
0 & \sqrt{1} & 0 & . & . & . \\
0 & 0 & \sqrt{2} & 0 & . & . \\
0 & 0 & 0 & \sqrt{3} & 0 & . \\
. & . & . & . & . & . \\
. & . & . & . & . & . \\
. & . & . & . & . & .
\end{pmatrix}
\begin{pmatrix}
0 & . & . & . & . & . \\
\sqrt{1} & 0 & . & . & . & . \\
0 & \sqrt{2} & 0 & . & . & . \\
0 & 0 & \sqrt{3} & 0 & . & . \\
. & . & . & . & . & . \\
. & . & . & . & . & .
\end{pmatrix}
$$

$$
=
\begin{pmatrix}
1 & 0 & . & . & . & . \\
0 & 2 & 0 & . & . & . \\
0 & 0 & 3 & 0 & . & . \\
. & . & . & . & . & . \\
. & . & . & . & . & . \\
. & . & . & . & . & .
\end{pmatrix}.
\tag{15.61}
$$

Similarly,

$$
\hat{a}^{\dagger}\hat{a} =
\begin{pmatrix}
0 & . & . & . & . \\
\sqrt{1} & 0 & . & . & . \\
0 & \sqrt{2} & 0 & . & . \\
0 & 0 & \sqrt{3} & 0 & . \\
. & . & . & . & . \\
. & . & . & . & .
\end{pmatrix}
\begin{pmatrix}
0 & \sqrt{1} & 0 & . & . & . \\
0 & 0 & \sqrt{2} & 0 & . & . \\
0 & 0 & 0 & \sqrt{3} & 0 & . \\
. & . & . & . & . & . \\
. & . & . & . & . & . \\
. & . & . & . & . & .
\end{pmatrix}
$$

$$
=
\begin{pmatrix}
0 & 0 & . & . & . & . \\
0 & 1 & 0 & . & . & . \\
0 & 0 & 2 & 0 & . & . \\
. & . & . & . & . & . \\
. & . & . & . & . & . \\
. & . & . & . & . & .
\end{pmatrix}.
\tag{15.62}
$$

Thus,

$$
[\hat{a}, \hat{a}^{\dagger}] = \hat{a}\hat{a}^{\dagger} - \hat{a}^{\dagger}\hat{a} =
\begin{pmatrix}
1 & 0 & . & . & . & . \\
0 & 1 & 0 & . & . & . \\
0 & 0 & 1 & 0 & . & . \\
. & . & . & . & . & . \\
. & . & . & . & . & . \\
. & . & . & . & . & .
\end{pmatrix}
= 1.
\tag{15.63}
$$

Problem 15.7

Introducing a dimensionless parameter $\xi = \sqrt{\frac{m\omega}{\hbar}}x$,

(a) Show that the operators $\hat{a}$ and $\hat{a}^\dagger$ can be written as

$$\hat{a} = \frac{1}{\sqrt{2}}\left(\xi + \frac{\partial}{\partial\xi}\right),$$

$$\hat{a}^\dagger = \frac{1}{\sqrt{2}}\left(\xi - \frac{\partial}{\partial\xi}\right). \tag{15.64}$$

(b) Show that the time-independent Schrödinger equation becomes

$$\frac{\partial^2\phi}{\partial\xi^2} + \left(\frac{2E}{\hbar\omega} - \xi^2\right)\phi = 0. \tag{15.65}$$

(c) Show that the wave function $\phi_1(x)$ of the $n = 1$ energy state can be written as

$$\phi_1(x) = 2\xi A_1 e^{-\xi^2/2}. \tag{15.66}$$

(d) Find the normalization constant A_1.

(e) Using (a) as the representation of the operators $\hat{a}$ and $\hat{a}^\dagger$, verify the commutation relation $\left[\hat{a}, \hat{a}^\dagger\right] = 1$.

Solution (a)

Using the relations

$$\hat{a} = \sqrt{\frac{m\omega}{2\hbar}}\,\hat{x} + i\frac{1}{\sqrt{2m\hbar\omega}}\,\hat{p},$$

$$\hat{a}^\dagger = \sqrt{\frac{m\omega}{2\hbar}}\,\hat{x} - i\frac{1}{\sqrt{2m\hbar\omega}}\,\hat{p}, \tag{15.67}$$

and the fact that

$$\xi = \sqrt{\frac{m\omega}{\hbar}}x, \tag{15.68}$$

and that

$$\hat{p} = -i\hbar\frac{\partial}{\partial x} = -i\hbar\frac{\partial}{\partial\xi}\frac{\partial\xi}{\partial x} = -i\hbar\sqrt{\frac{m\omega}{\hbar}}\frac{\partial}{\partial\xi}, \tag{15.69}$$

we obtain

$$\hat{a} = \frac{1}{\sqrt{2}}\xi + \hbar\frac{1}{\sqrt{2}}\sqrt{\frac{m\omega}{m\hbar^2\omega}}\frac{\partial}{\partial\xi} = \frac{1}{\sqrt{2}}\left(\xi + \frac{\partial}{\partial\xi}\right). \tag{15.70}$$

Similarly, we can show that

$$\hat{a}^\dagger = \frac{1}{\sqrt{2}}\left(\xi - \frac{\partial}{\partial\xi}\right). \tag{15.71}$$

Solution (b)

We start from the time-independent Schrödinger equation

$$\hat{H}\phi = E\phi, \tag{15.72}$$

where

$$\hat{H} = \hbar\omega\left(\hat{a}^{\dagger}\hat{a} + \frac{1}{2}\right). \tag{15.73}$$

Using the results from part (a)

$$\hat{a} = \frac{1}{\sqrt{2}}\left(\xi + \frac{\partial}{\partial\xi}\right),$$

$$\hat{a}^{\dagger} = \frac{1}{\sqrt{2}}\left(\xi - \frac{\partial}{\partial\xi}\right), \tag{15.74}$$

we have

$$
\begin{aligned}
\hat{a}^{\dagger}\hat{a}\phi &= \frac{1}{2}\left(\xi - \frac{\partial}{\partial\xi}\right)\left(\xi + \frac{\partial}{\partial\xi}\right)\phi = \frac{1}{2}\left(\xi - \frac{\partial}{\partial\xi}\right)\left(\xi\phi + \frac{\partial\phi}{\partial\xi}\right) \\
&= \frac{1}{2}\left(\xi^2\phi + \xi\frac{\partial\phi}{\partial\xi} - \frac{\partial}{\partial\xi}(\xi\phi) - \frac{\partial^2\phi}{\partial\xi^2}\right) \\
&= \frac{1}{2}\left(\xi^2\phi + \xi\frac{\partial\phi}{\partial\xi} - \phi - \xi\frac{\partial\phi}{\partial\xi} - \frac{\partial^2\phi}{\partial\xi^2}\right) \\
&= \frac{1}{2}\left(\xi^2\phi - \phi - \frac{\partial^2\phi}{\partial\xi^2}\right).
\end{aligned}
\tag{15.75}
$$

Hence, the time-independent Schrödinger can be written as

$$\hat{H}\phi - E\phi = \frac{1}{2}\hbar\omega\left(\xi^2\phi - \phi - \frac{\partial^2\phi}{\partial\xi^2} + \phi\right) - E\phi = 0, \tag{15.76}$$

which can be rewritten as

$$\frac{\partial^2\phi}{\partial\xi^2} - \xi^2\phi + \frac{2E}{\hbar\omega}\phi = 0, \tag{15.77}$$

and finally

$$\frac{\partial^2\phi}{\partial\xi^2} + \left(\frac{2E}{\hbar\omega} - \xi^2\right)\phi = 0. \tag{15.78}$$

Solution (c)

The wave function $\phi_1(x)$ is of the form (see Section 15.1 of the textbook, Eq. (15.41)):

$$\phi_1(x) = \sqrt{2}\sqrt{\frac{m\omega}{\hbar}}\, x\phi_0(x). \qquad (15.79)$$

Since

$$\xi = \sqrt{\frac{m\omega}{\hbar}}x \quad \text{and} \quad \phi_0(x) = \phi_0(0)e^{-\frac{m\omega}{2\hbar}x^2}, \qquad (15.80)$$

we have

$$\phi_1(x) = \sqrt{2}\phi_0(0)\xi e^{-\xi^2/2} = 2A_1\xi e^{-\xi^2/2}, \qquad (15.81)$$

where

$$A_1 = \frac{1}{\sqrt{2}}\phi_0(0). \qquad (15.82)$$

Solution (d)

From the normalization condition

$$\int_{-\infty}^{+\infty} |\phi_1(x)|^2\, dx = 1, \qquad (15.83)$$

and from part (c), we have

$$4|A_1|^2 \int_{-\infty}^{+\infty} \xi^2 e^{-\xi^2}\, dx = 1. \qquad (15.84)$$

However,

$$dx = \sqrt{\frac{\hbar}{m\omega}}\, d\xi, \qquad (15.85)$$

so we have the normalization condition of the form

$$4\sqrt{\frac{\hbar}{m\omega}}|A_1|^2 \int_{-\infty}^{+\infty} \xi^2 e^{-\xi^2}\, d\xi = 1. \qquad (15.86)$$

Since

$$\int_{-\infty}^{+\infty} \xi^2 e^{-\xi^2}\, d\xi = \frac{1}{2}\sqrt{\pi}, \qquad (15.87)$$

we have

$$4\sqrt{\frac{\hbar}{m\omega}}|A_1|^2\frac{1}{2}\sqrt{\pi} = 1, \qquad (15.88)$$

from which we find

$$|A_1| = \frac{1}{\sqrt{2}}\left(\frac{m\omega}{\pi\hbar}\right)^{\frac{1}{4}}. \qquad (15.89)$$

Solution (e)

Consider the action of the commutator on a wave function ϕ:

$$\left[\hat{a}, \hat{a}^\dagger\right]\phi. \tag{15.90}$$

From the definition of the commutator and part (a), we have

$$
\begin{aligned}
\left(\hat{a}\hat{a}^\dagger - \hat{a}^\dagger\hat{a}\right)\phi &= \frac{1}{2}\left[\left(\xi + \frac{\partial}{\partial\xi}\right)\left(\xi - \frac{\partial}{\partial\xi}\right) - \left(\xi - \frac{\partial}{\partial\xi}\right)\left(\xi + \frac{\partial}{\partial\xi}\right)\right]\phi \\
&= \frac{1}{2}\left[\left(\xi + \frac{\partial}{\partial\xi}\right)\left(\xi\phi - \frac{\partial\phi}{\partial\xi}\right)\right. \\
&\quad \left. - \left(\xi - \frac{\partial}{\partial\xi}\right)\left(\xi\phi + \frac{\partial\phi}{\partial\xi}\right)\right] \\
&= \frac{1}{2}\left[\xi^2\phi - \xi\frac{\partial\phi}{\partial\xi} + \frac{\partial}{\partial\xi}(\xi\phi) - \frac{\partial^2\phi}{\partial\xi^2}\right. \\
&\quad \left. -\xi^2\phi - \xi\frac{\partial\phi}{\partial\xi} + \frac{\partial}{\partial\xi}(\xi\phi) + \frac{\partial^2\phi}{\partial\xi^2}\right] \\
&= \frac{1}{2}\left[-2\xi\frac{\partial\phi}{\partial\xi} + 2\xi\frac{\partial\phi}{\partial\xi} + 2\phi\right] = \phi. \tag{15.91}
\end{aligned}
$$

Hence

$$\left[\hat{a}, \hat{a}^\dagger\right] = 1. \tag{15.92}$$

Problem 15.8

Calculate the expectation value $\langle\hat{x}\rangle$ and the variance (fluctuations) $\sigma = \langle\hat{x}^2\rangle - \langle\hat{x}\rangle^2$ of the position operator of a one-dimensional harmonic oscillator being in the ground state $\phi_0(x)$, using

(a) Integral definition of the average.
(b) Dirac notation, which allows to express $\hat{x}$ in terms of $\hat{a}$, $\hat{a}^\dagger$, and to apply the result of the Tutorial Problem 15.5.
(c) Show that the average values of the kinetic and potential energies of a one-dimensional harmonic oscillator in an energy eigenstate $|\phi_n\rangle$ are equal.

Solution (a)

The wave function of the position operator of a one-dimensional harmonic oscillator being in the ground state is of the form

$$\phi_0(x) = Ae^{-\beta x^2}, \tag{15.93}$$

where

$$\beta = \frac{m\omega}{2\hbar} \quad \text{and} \quad A = \left(\frac{m\omega}{\pi\hbar}\right)^{\frac{1}{4}} = \left(\frac{2\beta}{\pi}\right)^{\frac{1}{4}}. \tag{15.94}$$

Thus, the expectation value of the position operator is

$$\langle \hat{x} \rangle = \int_{-\infty}^{+\infty} \phi_0^*(x) x \phi_0(x)\, dx = |A|^2 \int_{-\infty}^{+\infty} x e^{-2\beta x^2}\, dx = 0, \tag{15.95}$$

since the function under the integral is an odd function.

Calculate now $\langle \hat{x}^2 \rangle$. From the definition of the expectation value, we have

$$\langle \hat{x}^2 \rangle = \int_{-\infty}^{+\infty} \phi_0^*(x) x^2 \phi_0(x)\, dx = |A|^2 \int_{-\infty}^{+\infty} x^2 e^{-2\beta x^2}\, dx. \tag{15.96}$$

Since

$$\int_{-\infty}^{+\infty} x^2 e^{-2\beta x^2}\, dx = \frac{1}{4\beta}\sqrt{\frac{\pi}{2\beta}}, \tag{15.97}$$

we have for the variance

$$\sigma = \langle \hat{x}^2 \rangle - \langle \hat{x} \rangle^2 = |A|^2 \frac{1}{4\beta}\sqrt{\frac{\pi}{2\beta}} - 0 = \left(\frac{2\beta}{\pi}\right)^{\frac{1}{2}} \frac{1}{4\beta}\sqrt{\frac{\pi}{2\beta}} = \frac{1}{4\beta}. \tag{15.98}$$

Solution (b)

Using the representation of $\hat{x}$ in terms of $\hat{a}$, $\hat{a}^\dagger$, we have

$$\hat{x} = \frac{1}{2\sqrt{\beta}}\left(\hat{a} + \hat{a}^\dagger\right). \tag{15.99}$$

Hence, we can write the expectation value of $\hat{x}$ as

$$\langle \hat{x} \rangle = \frac{1}{2\sqrt{\beta}}\langle \phi_0 | \hat{a} + \hat{a}^\dagger | \phi_0 \rangle = \frac{1}{2\sqrt{\beta}}\left(\langle \phi_0 | \hat{a} | \phi_0 \rangle + \langle \phi_0 | \hat{a}^\dagger | \phi_0 \rangle\right). \tag{15.100}$$

Since

$$\hat{a}|\phi_0\rangle = 0, \qquad \hat{a}^\dagger|\phi_0\rangle = |\phi_1\rangle, \qquad \text{and} \qquad \langle\phi_0|\phi_1\rangle = 0, \quad (15.101)$$

we have for the expectation value

$$\langle\hat{x}\rangle = 0. \qquad (15.102)$$

Calculate now $\langle\hat{x}^2\rangle$:

$$\langle\hat{x}^2\rangle = \frac{1}{4\beta}\langle\phi_0|(\hat{a} + \hat{a}^\dagger)(\hat{a} + \hat{a}^\dagger)|\phi_0\rangle$$

$$= \frac{1}{4\beta}\left(\langle\phi_0|\hat{a}\hat{a}|\phi_0\rangle + \langle\phi_0|\hat{a}\hat{a}^\dagger|\phi_0\rangle\right.$$

$$\left. + \langle\phi_0|\hat{a}^\dagger\hat{a}|\phi_0\rangle + \langle\phi_0|\hat{a}^\dagger\hat{a}^\dagger|\phi_0\rangle\right). \qquad (15.103)$$

Since

$$\hat{a}\hat{a}|\phi_0\rangle = 0,$$

$$\hat{a}^\dagger\hat{a}|\phi_0\rangle = 0,$$

$$\hat{a}\hat{a}^\dagger|\phi_0\rangle = \hat{a}|\phi_1\rangle = |\phi_0\rangle,$$

$$\hat{a}^\dagger\hat{a}^\dagger|\phi_0\rangle = \sqrt{2}|\phi_2\rangle, \qquad (15.104)$$

and $\langle\phi_0|\phi_2\rangle = 0$, we obtain

$$\langle\hat{x}^2\rangle = \frac{1}{4\beta}. \qquad (15.105)$$

Hence, the variance is

$$\sigma = \langle\hat{x}^2\rangle - \langle\hat{x}\rangle^2 = \frac{1}{4\beta}, \qquad (15.106)$$

which is the same value as predicted in part (a).

Solution (c)

The kinetic and potential energies of the harmonic oscillator are defined as

$$\hat{E}_k = \frac{1}{2m}\hat{p}^2, \qquad \hat{V} = \frac{1}{2}m\omega^2\hat{x}^2. \qquad (15.107)$$

Consider first the kinetic energy. Since

$$\hat{p} = -i\sqrt{\frac{m\omega\hbar}{2}}(\hat{a} - \hat{a}^\dagger), \qquad (15.108)$$

we have

$$\hat{p}^2 = -\frac{m\omega\hbar}{2} \left(\hat{a} - \hat{a}^\dagger \right) \left(\hat{a} - \hat{a}^\dagger \right).$$ (15.109)

Hence,

$$\begin{aligned}
\langle \hat{E}_k \rangle &= \langle \phi_n | \hat{E}_k | \phi_n \rangle \\
&= -\frac{1}{4}\hbar\omega \left(\langle \phi_n | \hat{a}\hat{a} | \phi_n \rangle - \langle \phi_n | \hat{a}\hat{a}^\dagger | \phi_n \rangle \right. \\
&\quad \left. - \langle \phi_n | \hat{a}^\dagger \hat{a} | \phi_n \rangle + \langle \phi_n | \hat{a}^\dagger \hat{a}^\dagger | \phi_n \rangle \right).
\end{aligned}$$ (15.110)

Since

$$\begin{aligned}
\hat{a}\hat{a}|\phi_n\rangle &= \sqrt{n(n-1)}|\phi_{n-2}\rangle, \\
\hat{a}^\dagger \hat{a}|\phi_n\rangle &= n|\phi_n\rangle, \\
\hat{a}\hat{a}^\dagger|\phi_n\rangle &= (n+1)|\phi_n\rangle, \\
\hat{a}^\dagger \hat{a}^\dagger|\phi_n\rangle &= \sqrt{(n+1)(n+2)}|\phi_{n+2}\rangle,
\end{aligned}$$ (15.111)

and $\langle \phi_n | \phi_m \rangle = \delta_{nm}$, we obtain

$$\langle \hat{E}_k \rangle = -\frac{1}{4}\hbar\omega[0 - (n+1) - n + 0] = \frac{1}{4}\hbar\omega(2n+1) = \frac{1}{2}\hbar\omega\left(n + \frac{1}{2}\right).$$ (15.112)

Consider now the expectation value of the potential energy. Since

$$\hat{x} = \frac{1}{2}\sqrt{\frac{2\hbar}{m\omega}} \left(\hat{a} + \hat{a}^\dagger \right),$$ (15.113)

we have

$$\begin{aligned}
\langle \hat{V} \rangle &= \langle \phi_n | \hat{V} | \phi_n \rangle = \frac{1}{2}m\omega^2 \frac{2\hbar}{4m\omega} \langle \phi_n | \left(\hat{a} + \hat{a}^\dagger \right)^2 | \phi_n \rangle \\
&= \frac{1}{4}\hbar\omega \left(\langle \phi_n | \hat{a}\hat{a} | \phi_n \rangle + \langle \phi_n | \hat{a}\hat{a}^\dagger | \phi_n \rangle \right. \\
&\quad \left. + \langle \phi_n | \hat{a}^\dagger \hat{a} | \phi_n \rangle + \langle \phi_n | \hat{a}^\dagger \hat{a}^\dagger | \phi_n \rangle \right).
\end{aligned}$$ (15.114)

Hence,

$$\langle \hat{V} \rangle = \frac{1}{4}\hbar\omega[0 + (n+1) + n + 0] = \frac{1}{4}\hbar\omega(2n+1) = \frac{1}{2}\hbar\omega\left(n + \frac{1}{2}\right).$$ (15.115)

Thus,

$$\langle \hat{E}_k \rangle = \langle \hat{V} \rangle.$$ (15.116)

Problem 15.9

Show that the non-zero minimum energy of the quantum harmonic oscillator, $E \geq \hbar\omega/2$, is the consequence of the uncertainty relation between the position and momentum operators of the oscillator.

(*Hint:* Use the uncertainty relation for the position and momentum operators in the state $n = 0$ and plug it into the average energy of the oscillator. Then, find the minimum of the energy with respect to δx.)

Solution

Take the square of the uncertainty relation for δx and δp in the state $n = 0$:

$$\delta x^2 \delta p^2 \geq \frac{\hbar^2}{4}, \qquad (15.117)$$

and the expression for the average energy of the harmonic oscillator

$$\langle E \rangle = \frac{1}{2m}\langle \hat{p}^2 \rangle + \frac{1}{2}m\omega^2\langle \hat{x}^2 \rangle. \qquad (15.118)$$

Since $\langle \hat{x} \rangle = 0$ and $\langle \hat{p} \rangle = 0$, we have $\delta x^2 = \langle \hat{x}^2 \rangle$ and $\delta p^2 = \langle \hat{p}^2 \rangle$, so that we can write

$$\langle E \rangle = \frac{1}{2m}\delta p^2 + \frac{1}{2}m\omega^2\delta x^2. \qquad (15.119)$$

From this expression, we have

$$\delta p^2 = 2m\langle E \rangle - m^2\omega^2\delta x^2, \qquad (15.120)$$

and substituting it into Eq. (15.117), we get

$$\delta x^2 \left(2m\langle E \rangle - m^2\omega^2\delta x^2 \right) \geq \frac{\hbar^2}{4}. \qquad (15.121)$$

From this, we find

$$\langle E \rangle \geq \frac{\hbar^2}{8m\delta x^2} + \frac{1}{2}m\omega^2\delta x^2. \qquad (15.122)$$

Since

$$\delta x^2 = \frac{1}{2}\frac{\hbar}{m\omega}, \qquad (15.123)$$

we have

$$\langle E \rangle \geq \frac{\hbar\omega}{4} + \frac{\hbar\omega}{4} = \frac{\hbar\omega}{2}. \tag{15.124}$$

Thus, starting from the uncertainty relation for the position and momentum operators of the quantum harmonic oscillator, the minimum energy of the oscillator is $\hbar\omega/2$.

Chapter 16

Quantum Theory of Hydrogen Atom

Problem 16.2

The normalized wave function of the ground state of a hydrogen-like atom with nuclear charge Ze has the form

$$|\Psi\rangle = Ae^{-\beta r}, \qquad (16.1)$$

where A and β are real constants, and r is the distance between the electron and the nucleus. The Hamiltonian of the atom is given by

$$\hat{H} = -\frac{\hbar^2}{2m}\nabla^2 - \frac{Ze^2}{4\pi\varepsilon_0}\frac{1}{r}. \qquad (16.2)$$

Show that

(a) $A^2 = \beta^3/\pi$.
(b) $\beta = Z/a_o$, where a_o is the Bohr radius.
(c) The energy of the electron is $E = -Z^2 E_0$, where $E_0 = e^2/(8\pi\varepsilon_0 a_o)$.
(d) The expectation values of the potential and kinetic energies are $2E$ and $-E$, respectively.

Problems and Solutions In Quantum Physics
Zbigniew Ficek
Copyright © 2016 Pan Stanford Publishing Pte. Ltd.
ISBN 978-981-4669-36-8 (Hardcover), 978-981-4669-37-5 (eBook)
www.panstanford.com

Solution (a)

The constant A is found from the normalization condition

$$1 = \langle \Psi | \Psi \rangle \equiv \int |\Psi|^2 dV = 4\pi A^2 \int_0^\infty r^2 e^{-2\beta r} dr$$

$$= 4\pi A^2 \frac{2}{(2\beta)^3} = \frac{\pi A^2}{\beta^3}. \qquad (16.3)$$

Hence, $A^2 = \beta^3 / \pi$.

Solution (b)

We find β from the condition that the wave function $|\Psi\rangle$ is a solution to the stationary Schrödinger equation for a hydrogen-like atom

$$-\frac{\hbar^2}{2m} \nabla^2 |\Psi\rangle + V(r)|\Psi\rangle = E|\Psi\rangle. \qquad (16.4)$$

We see that we have to evaluate $\nabla^2 |\Psi\rangle$. Since the wave function is given in the spherical coordinates, we have

$$\nabla^2 |\Psi\rangle = A^2 \nabla^2 e^{-\beta r} = A^2 \left(\frac{\partial^2}{\partial r^2} + \frac{2}{r} \frac{\partial}{\partial r} \right) e^{-\beta r}$$

$$= A^2 \left(\beta^2 - \frac{2\beta}{r} \right) e^{-\beta r} = \left(\beta^2 - \frac{2\beta}{r} \right) |\Psi\rangle. \qquad (16.5)$$

Hence, the Schrödinger equation (16.4) takes the form

$$\left[-\frac{\hbar^2}{2m} \left(\beta^2 - \frac{2\beta}{r} \right) + V(r) - E \right] |\Psi\rangle = 0, \qquad (16.6)$$

which after substituting, the explicit form of $V(x)$ reduces to

$$\left[-\frac{\hbar^2}{2m} \left(\beta^2 - \frac{2\beta}{r} \right) - \frac{\alpha}{r} - E \right] |\Psi\rangle = 0, \qquad (16.7)$$

where $\alpha = Ze^2 / (4\pi \varepsilon_0)$.

Since $|\Psi\rangle$ is different from zero, the left-hand side of Eq. (16.7) will be zero only if

$$-\frac{\hbar^2}{2m} \left(\beta^2 - \frac{2\beta}{r} \right) - \frac{\alpha}{r} - E = 0. \qquad (16.8)$$

We know that the energy E of the electron in a given energy state of the hydrogen atom is a constant independent of r. Therefore, the terms dependent on r in Eq. (16.8) must add to zero. This happens when

$$\frac{\hbar^2 \beta}{m} - \alpha = 0,\qquad(16.9)$$

which gives

$$\beta = \frac{m}{\hbar^2}\alpha = \frac{m}{\hbar^2}\frac{Ze^2}{4\pi\varepsilon_0} = \frac{Z}{a_o}.\qquad(16.10)$$

Solution (c)

Since the terms dependent on r in Eq. (16.8) are equal to zero, we have

$$E = -\frac{\hbar^2}{2m}\beta^2 = -Z^2\frac{me^4}{2\hbar^2(4\pi\varepsilon_0)^2} = -Z^2 E_0,\qquad(16.11)$$

where $E_0 = e^2/(8\pi\varepsilon_0 a_o)$.

Solution (d)

The expectation value of the potential energy is

$$\langle V(r)\rangle = \int \Psi^* V(r)\Psi dV = -4\pi A^2\alpha\int_0^\infty r e^{-2\beta r}dr$$

$$= -4\alpha\beta^3\frac{1}{(2\beta)^2} = -\alpha\beta = -\frac{\hbar^2}{m}\beta^2 = 2E.\qquad(16.12)$$

The expectation value of the kinetic energy is

$$\langle E_k\rangle = \int \Psi^* E_k\Psi dV = -4\pi A^2\frac{\hbar^2}{2m}\int_0^\infty r^2 e^{-\beta r}\nabla^2 e^{-\beta r}dr$$

$$= -4\pi A^2\frac{\hbar^2}{2m}\int_0^\infty r^2 e^{-\beta r}\left(\beta^2 - \frac{2\beta}{r}\right)e^{-\beta r}dr$$

$$= -4\pi A^2\frac{\hbar^2}{2m}\left[\beta^2\int_0^\infty r^2 e^{-2\beta r}dr - 2\beta\int_0^\infty r e^{-2\beta r}dr\right]$$

$$= -4\pi A^2\frac{\hbar^2}{2m}\left[\beta^2\frac{2}{(2\beta)^3} - 2\beta\frac{1}{(2\beta)^2}\right]$$

$$= -2\beta^3\frac{\hbar^2}{m}\left(\frac{1}{4\beta} - \frac{1}{2\beta}\right) = \frac{1}{2}\beta^2\frac{\hbar^2}{m} = \frac{\hbar^2}{2m}\beta^2 = -E.\qquad(16.13)$$

We see that

$$\langle E_k \rangle = -\frac{1}{2} \langle V(r) \rangle. \tag{16.14}$$

This result is a special case of the *virial theorem*, which states that for a system in a stationary state in a potential $V(r)$ proportional to r^n,

$$\langle E_k \rangle = \frac{n}{2} \langle V(r) \rangle. \tag{16.15}$$

Thus, for the electron in a hydrogen-like atom, the potential is inversely proportional to r ($n = -1$), which gives the result in Eq. (16.14). The virial theorem holds also in classical physics, and as we know from classical mechanics, the result (16.14) applies, e.g., to a satellite orbiting the Earth.

Problem 16.3

Consider the angular momentum operator $\hat{L} = \hat{r} \times \hat{p}$. Show that

(a) The operator $\hat{L}$ is Hermitian.
   (Hint: Show that the components L_x, L_y, L_z are Hermitian).
(b) The components of $\hat{L}$ (L_x, L_y, L_z) do not commute.
(c) The square of the angular momentum $\hat{L}^2$ commutes with each of the components L_x, L_y, L_z.
(d) In the spherical coordinates, the components and the square of the angular momentum can be expressed as

$$L_x = -i\hbar \left(-\sin\phi \frac{\partial}{\partial\theta} - \frac{\cos\phi \cos\theta}{\sin\theta} \frac{\partial}{\partial\phi} \right),$$

$$L_y = -i\hbar \left(\cos\phi \frac{\partial}{\partial\theta} - \frac{\sin\phi \cos\theta}{\sin\theta} \frac{\partial}{\partial\phi} \right),$$

$$L_z = -i\hbar \frac{\partial}{\partial\phi},$$

$$L^2 = -\hbar^2 \left[\frac{1}{\sin^2\theta} \frac{\partial^2}{\partial\phi^2} + \frac{1}{\sin\theta} \frac{\partial}{\partial\theta} \left(\sin\theta \frac{\partial}{\partial\theta} \right) \right]. \tag{16.16}$$

Solution (a)

The operator $\hat{\vec{L}}$ is Hermitian if the components $\hat{L}_x, \hat{L}_y, \hat{L}_z$ are Hermitian, i.e., if

$$\hat{L}_x^\dagger = \hat{L}_x, \qquad \hat{L}_y^\dagger = \hat{L}_y, \qquad \hat{L}_z^\dagger = \hat{L}_z. \tag{16.17}$$

First, we find the components $\hat{L}_x, \hat{L}_y, \hat{L}_z$ in terms of the position and momentum operators, which are Hermitian. Since the angular momentum operator can be written as

$$\hat{\vec{L}} = \hat{L}_x \vec{i} + \hat{L}_y \vec{j} + \hat{L}_z \vec{k} = \hat{\vec{r}} \times \hat{\vec{p}}$$
$$= \left(\hat{y}\hat{p}_z - \hat{z}\hat{p}_y \right) \vec{i} + \left(\hat{z}\hat{p}_x - \hat{x}\hat{p}_z \right) \vec{j} + \left(\hat{x}\hat{p}_y - \hat{y}\hat{p}_x \right) \vec{k}, \tag{16.18}$$

we find that

$$\hat{L}_x = \left(\hat{y}\hat{p}_z - \hat{z}\hat{p}_y \right), \quad \hat{L}_y = \left(\hat{z}\hat{p}_x - \hat{x}\hat{p}_z \right), \quad \hat{L}_z = \left(\hat{x}\hat{p}_y - \hat{y}\hat{p}_x \right). \tag{16.19}$$

Consider $\hat{L}_x$:

$$\hat{L}_x^\dagger = \left(\hat{y}\hat{p}_z - \hat{z}\hat{p}_y \right)^\dagger = \left(\hat{y}\hat{p}_z \right)^\dagger - \left(\hat{z}\hat{p}_y \right)^\dagger = \hat{p}_z^\dagger \hat{y}^\dagger - \hat{p}_y^\dagger \hat{z}^\dagger = \hat{p}_z \hat{y} - \hat{p}_y \hat{z}. \tag{16.20}$$

Since

$$[\hat{y}, \hat{p}_z] = [\hat{z}, \hat{p}_y] = 0, \tag{16.21}$$

we find that

$$\hat{L}_x^\dagger = \hat{p}_z \hat{y} - \hat{p}_y \hat{z} = \hat{y}\hat{p}_z - \hat{z}\hat{p}_y = \hat{L}_x. \tag{16.22}$$

Similarly, we can show that $\hat{L}_y^\dagger = \hat{L}_y$ and $\hat{L}_z^\dagger = \hat{L}_z$. Hence, $\hat{\vec{L}}$ is Hermitian.

Solution (b)

Consider a commutator

$$\left[\hat{L}_x, \hat{L}_y \right]. \tag{16.23}$$

Using the expressions (16.19), we obtain

$$\left[\hat{L}_x, \hat{L}_y \right] = \left(\hat{y}\hat{p}_z - \hat{z}\hat{p}_y \right) \left(\hat{z}\hat{p}_x - \hat{x}\hat{p}_z \right) - \left(\hat{z}\hat{p}_x - \hat{x}\hat{p}_z \right) \left(\hat{y}\hat{p}_z - \hat{z}\hat{p}_y \right)$$
$$= \hat{y}\hat{p}_z\hat{z}\hat{p}_x - \hat{y}\hat{p}_z\hat{x}\hat{p}_z - \hat{z}\hat{p}_y\hat{z}\hat{p}_x + \hat{z}\hat{p}_y\hat{x}\hat{p}_z$$
$$\quad - \hat{z}\hat{p}_x\hat{y}\hat{p}_z + \hat{z}\hat{p}_x\hat{z}\hat{p}_y + \hat{x}\hat{p}_z\hat{y}\hat{p}_z - \hat{x}\hat{p}_z\hat{z}\hat{p}_y. \tag{16.24}$$

Since

$$\left[\hat{y},\ \hat{p}_z\right] = \left[\hat{z},\ \hat{p}_y\right] = \left[\hat{x},\ \hat{p}_z\right] = \left[\hat{z},\ \hat{p}_x\right] = 0, \qquad (16.25)$$

we have

$$\hat{y}\hat{p}_z\hat{x}\hat{p}_z = \hat{x}\hat{p}_z\hat{y}\hat{p}_z \qquad \text{and} \qquad \hat{z}\hat{p}_y\hat{z}\hat{p}_x = \hat{z}\hat{p}_x\hat{z}\hat{p}_y. \qquad (16.26)$$

Thus,

$$\left[\hat{L}_x,\ \hat{L}_y\right] = \hat{y}\hat{p}_z\hat{z}\hat{p}_x + \hat{z}\hat{p}_y\hat{x}\hat{p}_z - \hat{z}\hat{p}_x\hat{y}\hat{p}_z - \hat{x}\hat{p}_z\hat{z}\hat{p}_y. \qquad (16.27)$$

Since $[\hat{z},\ \hat{p}_z] = i\hbar$, we can replace $\hat{p}_z\hat{z}$ by $\hat{z}\hat{p}_z - i\hbar$ and obtain

$$\left[\hat{L}_x,\ \hat{L}_y\right] = \hat{y}(\hat{z}\hat{p}_z - i\hbar)\hat{p}_x + \hat{z}\hat{p}_y\hat{x}\hat{p}_z - \hat{z}\hat{p}_x\hat{y}\hat{p}_z - \hat{x}(\hat{z}\hat{p}_z - i\hbar)\hat{p}_y$$

$$= i\hbar\left(\hat{x}\hat{p}_y - \hat{y}\hat{p}_x\right) = i\hbar\hat{L}_z. \qquad (16.28)$$

Consequently,

$$\left[\hat{L}_x,\ \hat{L}_y\right] = i\hbar\hat{L}_z. \qquad (16.29)$$

Similarly, we can show that

$$\left[\hat{L}_y,\ \hat{L}_z\right] = i\hbar\hat{L}_x \quad \text{and} \quad \left[\hat{L}_z,\ \hat{L}_x\right] = i\hbar\hat{L}_y. \qquad (16.30)$$

Solution (c)

Since

$$\hat{L}^2 = \hat{L}_x^2 + \hat{L}_y^2 + \hat{L}_z^2, \qquad (16.31)$$

we find that

$$\left[\hat{L}^2,\ \hat{L}_x\right] = \left[\hat{L}_x^2,\ \hat{L}_x\right] + \left[\hat{L}_y^2,\ \hat{L}_x\right] + \left[\hat{L}_z^2,\ \hat{L}_x\right] = \left[\hat{L}_y^2,\ \hat{L}_x\right] + \left[\hat{L}_z^2,\ \hat{L}_x\right]. \qquad (16.32)$$

Thus,

$$\left[\hat{L}^2,\ \hat{L}_x\right] = \hat{L}_y^2\hat{L}_x - \hat{L}_x\hat{L}_y^2 + \hat{L}_z^2\hat{L}_x - \hat{L}_x\hat{L}_z^2. \qquad (16.33)$$

Using the commutation relations of (b), we then find

$$\left[\hat{L}^2,\ \hat{L}_x\right] = \hat{L}_y^2\hat{L}_x - \hat{L}_x\hat{L}_y^2 + \hat{L}_z^2\hat{L}_x - \hat{L}_x\hat{L}_z^2$$

$$= \hat{L}_y\left(\hat{L}_x\hat{L}_y - i\hbar\hat{L}_z\right) - \hat{L}_x\hat{L}_y^2 + \hat{L}_z\left(\hat{L}_x\hat{L}_z + i\hbar\hat{L}_y\right) - \hat{L}_x\hat{L}_z^2$$

$$= (\hat{L}_y\hat{L}_x)\hat{L}_y - i\hbar\hat{L}_y\hat{L}_z - \hat{L}_x\hat{L}_y^2 + (\hat{L}_z\hat{L}_x)\hat{L}_z - \hat{L}_x\hat{L}_z^2$$

$$= \left(\hat{L}_x\hat{L}_y - i\hbar\hat{L}_z\right)\hat{L}_y - i\hbar\hat{L}_y\hat{L}_z - \hat{L}_x\hat{L}_y^2 + \left(\hat{L}_x\hat{L}_z + i\hbar\hat{L}_y\right)\hat{L}_z$$

$$\quad + i\hbar\hat{L}_z\hat{L}_y - \hat{L}_x\hat{L}_z^2$$

$$= 0. \qquad (16.34)$$

Similarly, we can show that

$$\left[\hat{L}^2,\ \hat{L}_y\right] = \left[\hat{L}^2,\ \hat{L}_z\right] = 0. \qquad (16.35)$$

Solution (d)

In spherical coordinates

$$x = r \sin\theta \cos\phi,$$
$$y = r \sin\theta \sin\phi,$$
$$z = r \cos\theta, \qquad\qquad (16.36)$$

and $r = \sqrt{x^2 + y^2 + z^2}$.

Consider $\hat{L}_x$, which in cartesian coordinates can be written as

$$\hat{L}_x = (\hat{y}\hat{p}_z - \hat{z}\hat{p}_y) = -i\hbar\left(y\frac{\partial}{\partial z} - z\frac{\partial}{\partial y}\right). \qquad (16.37)$$

Using the chain rule, we can express the derivatives in terms of the spherical components as

$$\frac{\partial}{\partial z} = \frac{\partial}{\partial r}\frac{\partial r}{\partial z} + \frac{\partial}{\partial\theta}\frac{\partial\theta}{\partial z} + \frac{\partial}{\partial\phi}\frac{\partial\phi}{\partial z},$$
$$\frac{\partial}{\partial y} = \frac{\partial}{\partial r}\frac{\partial r}{\partial y} + \frac{\partial}{\partial\theta}\frac{\partial\theta}{\partial y} + \frac{\partial}{\partial\phi}\frac{\partial\phi}{\partial y}. \qquad (16.38)$$

Since

$$r = \sqrt{x^2 + y^2 + z^2}, \quad \theta = \arccos\frac{z}{r}, \quad \phi = \arctan\frac{y}{x}, \qquad (16.39)$$

we find that

$$\frac{\partial r}{\partial z} = \frac{z}{r} = \cos\theta, \qquad \frac{\partial\theta}{\partial z} = -\frac{\sqrt{x^2+y^2}}{r} = -\frac{1}{r}\sin\theta, \qquad \frac{\partial\phi}{\partial z} = 0,$$

$$\frac{\partial r}{\partial y} = \frac{y}{r} = \sin\theta\sin\phi, \qquad \frac{\partial\theta}{\partial y} = \frac{yz}{\sqrt{r^2-z^2}} = \frac{1}{r}\cos\theta\sin\phi,$$

$$\frac{\partial\phi}{\partial y} = \frac{x}{x^2+y^2} = \frac{1}{r}\frac{\cos\phi}{\sin\theta},$$

$$\frac{\partial r}{\partial x} = \frac{x}{r} = \sin\theta\cos\phi, \qquad \frac{\partial\theta}{\partial x} = \frac{xz}{\sqrt{r^2-z^2}} = \frac{1}{r}\cos\theta\cos\phi,$$

$$\frac{\partial\phi}{\partial x} = -\frac{y}{x^2+y^2} = -\frac{1}{r}\frac{\sin\phi}{\sin\theta}. \qquad (16.40)$$

Consequently,

$$\hat{L}_x/(-i\hbar) = y\frac{\partial}{\partial z} - z\frac{\partial}{\partial y} = r\sin\theta\sin\phi\left(\cos\theta\frac{\partial}{\partial r} - \frac{1}{r}\sin\theta\frac{\partial}{\partial\theta}\right)$$

$$- r\cos\theta\left(\sin\theta\sin\phi\frac{\partial}{\partial r} + \frac{1}{r}\cos\theta\sin\phi\frac{\partial}{\partial\theta} + \frac{1}{r}\frac{\cos\phi}{\sin\theta}\frac{\partial}{\partial\phi}\right)$$

$$= -\left(\sin^2\theta + \cos^2\theta\right)\sin\phi\frac{\partial}{\partial\theta} - \frac{\cos\theta\cos\phi}{\sin\theta}\frac{\partial}{\partial\phi}$$

$$= -\sin\phi\frac{\partial}{\partial\theta} - \frac{\cos\theta\cos\phi}{\sin\theta}\frac{\partial}{\partial\phi}. \qquad (16.41)$$

Similarly,

$$\hat{L}_y/(-i\hbar) = z\frac{\partial}{\partial x} - x\frac{\partial}{\partial z}$$

$$= r\cos\theta\left(\sin\theta\cos\phi\frac{\partial}{\partial r} + \frac{1}{r}\cos\theta\cos\phi\frac{\partial}{\partial\theta} - \frac{1}{r}\frac{\sin\phi}{\sin\theta}\frac{\partial}{\partial\phi}\right)$$

$$-r\sin\theta\cos\phi\left(\cos\theta\frac{\partial}{\partial r} - \frac{1}{r}\sin\theta\frac{\partial}{\partial\theta}\right)$$

$$= \left(\sin^2\theta + \cos^2\theta\right)\cos\phi\frac{\partial}{\partial\theta} - \frac{\cos\theta\sin\phi}{\sin\theta}\frac{\partial}{\partial\phi}$$

$$= \cos\phi\frac{\partial}{\partial\theta} - \frac{\cos\theta\sin\phi}{\sin\theta}\frac{\partial}{\partial\phi}, \tag{16.42}$$

and

$$\hat{L}_z/(-i\hbar) = x\frac{\partial}{\partial y} - y\frac{\partial}{\partial x}$$

$$= r\sin\theta\cos\phi\left(\sin\theta\sin\phi\frac{\partial}{\partial r} + \frac{1}{r}\cos\theta\sin\phi\frac{\partial}{\partial\theta} + \frac{1}{r}\frac{\cos\phi}{\sin\theta}\frac{\partial}{\partial\phi}\right)$$

$$-r\sin\theta\sin\phi\left(\sin\theta\cos\phi\frac{\partial}{\partial r} + \frac{1}{r}\cos\theta\cos\phi\frac{\partial}{\partial\theta} - \frac{1}{r}\frac{\sin\phi}{\sin\theta}\frac{\partial}{\partial\phi}\right)$$

$$= \left(\sin^2\phi + \cos^2\phi\right)\frac{\partial}{\partial\phi} = \frac{\partial}{\partial\phi}. $$

$$\tag{16.43}$$

Having the angular momentum components L_x, L_y, and L_z in the spherical coordinates, we can find L^2 in the spherical coordinates:

$$\hat{L}^2 = \hat{L}_x^2 + \hat{L}_y^2 + \hat{L}_z^2. \tag{16.44}$$

Calculate $\hat{L}_x^2$:

$$\hat{L}_x^2/(-\hbar^2) = \left(-\sin\phi\frac{\partial}{\partial\theta} - \frac{\cos\theta\cos\phi}{\sin\theta}\frac{\partial}{\partial\phi}\right)$$

$$\times\left(-\sin\phi\frac{\partial}{\partial\theta} - \frac{\cos\theta\cos\phi}{\sin\theta}\frac{\partial}{\partial\phi}\right)$$

$$= \sin^2\phi\frac{\partial^2}{\partial\theta^2} + \sin\phi\cos\phi\frac{\partial}{\partial\theta}\frac{\cos\theta}{\sin\theta}\frac{\partial}{\partial\phi}$$

$$+ \frac{\cos\theta\cos\phi}{\sin\theta}\frac{\partial}{\partial\phi}\sin\phi\frac{\partial}{\partial\theta} + \frac{\cos^2\theta\cos\phi}{\sin^2\theta}\frac{\partial}{\partial\phi}\cos\phi\frac{\partial}{\partial\phi}.$$

$$\tag{16.45}$$

Next we calculate $\hat{L}_y^2$:

$$\hat{L}_y^2/(-\hbar^2) = \left(\cos\phi\frac{\partial}{\partial\theta} - \frac{\cos\theta\sin\phi}{\sin\theta}\frac{\partial}{\partial\phi}\right)$$

$$\times\left(\cos\phi\frac{\partial}{\partial\theta} - \frac{\cos\theta\sin\phi}{\sin\theta}\frac{\partial}{\partial\phi}\right)$$

$$= \cos^2\phi\frac{\partial^2}{\partial\theta^2} - \sin\phi\cos\phi\frac{\partial}{\partial\theta}\frac{\cos\theta}{\sin\theta}\frac{\partial}{\partial\phi}$$

$$-\frac{\cos\theta\sin\phi}{\sin\theta}\frac{\partial}{\partial\phi}\cos\phi\frac{\partial}{\partial\theta} + \frac{\cos^2\theta\sin\phi}{\sin^2\theta}\frac{\partial}{\partial\phi}\sin\phi\frac{\partial}{\partial\phi},$$

$$(16.46)$$

and $\hat{L}_z^2$:

$$\hat{L}_x^2/(-\hbar^2) = \frac{\partial^2}{\partial\phi^2}. \qquad (16.47)$$

Hence,

$$\hat{L}^2/(-\hbar^2) = \hat{L}_x^2/(-\hbar^2) + \hat{L}_y^2/(-\hbar^2) + \hat{L}_z^2/(-\hbar^2)$$

$$= (\sin^2\phi + \cos^2\phi)\frac{\partial^2}{\partial\theta^2}$$

$$+\frac{\cos\theta}{\sin\theta}\left(\cos\phi\frac{\partial}{\partial\phi}\sin\phi\frac{\partial}{\partial\theta} - \sin\phi\frac{\partial}{\partial\phi}\cos\phi\frac{\partial}{\partial\theta}\right)$$

$$+\frac{\cos^2\theta}{\sin^2\theta}\left(\cos\phi\frac{\partial}{\partial\phi}\cos\phi\frac{\partial}{\partial\phi} + \sin\phi\frac{\partial}{\partial\phi}\sin\phi\frac{\partial}{\partial\phi}\right) + \frac{\partial^2}{\partial\phi^2}$$

$$= \frac{\partial^2}{\partial\theta^2} + \frac{\cos\theta}{\sin\theta}\frac{\partial}{\partial\theta} + \frac{\cos^2\theta}{\sin^2\theta}\frac{\partial^2}{\partial\phi^2} + \frac{\partial^2}{\partial\phi^2}$$

$$= \left(\frac{\cos^2\theta}{\sin^2\theta} + 1\right)\frac{\partial^2}{\partial\phi^2} + \frac{1}{\sin\theta}\frac{\partial}{\partial\theta}\left(\sin\theta\frac{\partial}{\partial\theta}\right)$$

$$= \frac{1}{\sin^2\theta}\frac{\partial^2}{\partial\phi^2} + \frac{1}{\sin\theta}\frac{\partial}{\partial\theta}\left(\sin\theta\frac{\partial}{\partial\theta}\right). \qquad (16.48)$$

Problem 16.4

Particle in a potential of central symmetry

A particle of mass m moves in a potential of central symmetry, i.e., $V(x, y, z) = V(r)$. The energy of the particle is given by the Hamiltonian

$$\hat{H} = -\frac{\hbar^2}{2m}\nabla^2 + \hat{V}(r). \tag{16.49}$$

Show that $\hat{H}$ commutes with the angular momentum $\hat{\vec{L}}$ of the particle.

Solution

The angular momentum of the particle can be written as

$$\hat{\vec{L}} = \hat{L}_x \vec{i} + \hat{L}_y \vec{j} + \hat{L}_z \vec{k}, \tag{16.50}$$

and then the commutator splits into three commutators

$$\left[\hat{H}, \hat{\vec{L}}\right] = \left[\hat{H}, \hat{L}_x\right]\vec{i} + \left[\hat{H}, \hat{L}_y\right]\vec{j} + \left[\hat{H}, \hat{L}_z\right]\vec{k}. \tag{16.51}$$

Thus, $\hat{H}$ commutes with $\hat{\vec{L}}$ when $\hat{H}$ commutes with the components $\hat{L}_x$, $\hat{L}_y$, and $\hat{L}_z$.

Consider the commutator $\left[\hat{H}, \hat{L}_x\right]$, which can be written as the sum of two commutators

$$\left[\hat{H}, \hat{L}_x\right] = \left[-\frac{\hbar^2}{2m}\nabla^2 + \hat{V}(r), \hat{L}_x\right] = \left[-\frac{\hbar^2}{2m}\nabla^2, \hat{L}_x\right] + \left[\hat{V}(r), \hat{L}_x\right]. \tag{16.52}$$

First, we consider the commutator:

$$\left[-\frac{\hbar^2}{2m}\nabla^2, \hat{L}_x\right] = -\frac{\hbar^2}{2m}\left[\nabla^2, \hat{L}_x\right]. \tag{16.53}$$

Since

$$\hat{L}_x = -i\hbar\left(y\frac{\partial}{\partial z} - z\frac{\partial}{\partial y}\right), \tag{16.54}$$

we find

$$\left[-\frac{\hbar^2}{2m}\nabla^2, \hat{L}_x\right] = -\frac{\hbar^2}{2m}\left[\nabla^2, \hat{L}_x\right] = i\frac{\hbar^3}{2m}\left[\nabla^2, \left(y\frac{\partial}{\partial z} - z\frac{\partial}{\partial y}\right)\right]$$

$$= i\frac{\hbar^3}{2m}\left[\nabla^2\left(y\frac{\partial}{\partial z} - z\frac{\partial}{\partial y}\right) - \left(y\frac{\partial}{\partial z} - z\frac{\partial}{\partial y}\right)\nabla^2\right]$$

$$= i\frac{\hbar^3}{2m}\left[\left(\frac{\partial^2}{\partial x^2} + \frac{\partial^2}{\partial y^2} + \frac{\partial^2}{\partial z^2}\right)\left(y\frac{\partial}{\partial z} - z\frac{\partial}{\partial y}\right)\right.$$
$$\left. - \left(y\frac{\partial}{\partial z} - z\frac{\partial}{\partial y}\right)\left(\frac{\partial^2}{\partial x^2} + \frac{\partial^2}{\partial y^2} + \frac{\partial^2}{\partial z^2}\right)\right]$$

$$= i\frac{\hbar^3}{2m}\left[y\frac{\partial^2}{\partial x^2}\frac{\partial}{\partial z} - z\frac{\partial^2}{\partial x^2}\frac{\partial}{\partial y} + y\frac{\partial^2}{\partial y^2}\frac{\partial}{\partial z} - z\frac{\partial^3}{\partial y^3} + y\frac{\partial^3}{\partial z^3}\right.$$
$$- z\frac{\partial^2}{\partial z^2}\frac{\partial}{\partial y} - y\frac{\partial}{\partial z}\frac{\partial^2}{\partial x^2} - y\frac{\partial}{\partial z}\frac{\partial^2}{\partial y^2} - y\frac{\partial^3}{\partial z^3} + z\frac{\partial}{\partial y}\frac{\partial^2}{\partial x^2}$$
$$\left. + z\frac{\partial^3}{\partial y^3} + z\frac{\partial}{\partial y}\frac{\partial^2}{\partial z^2}\right]$$

$$= i\frac{\hbar^3}{2m}\left[y\left(\frac{\partial^2}{\partial x^2}\frac{\partial}{\partial z} - \frac{\partial}{\partial z}\frac{\partial^2}{\partial x^2}\right) + z\left(\frac{\partial}{\partial y}\frac{\partial^2}{\partial x^2} - \frac{\partial^2}{\partial x^2}\frac{\partial}{\partial y}\right)\right.$$
$$\left. + y\left(\frac{\partial^2}{\partial y^2}\frac{\partial}{\partial z} - \frac{\partial}{\partial z}\frac{\partial^2}{\partial y^2}\right) - z\left(\frac{\partial^2}{\partial z^2}\frac{\partial}{\partial y} - \frac{\partial}{\partial y}\frac{\partial^2}{\partial z^2}\right)\right].$$

$$(16.55)$$

Since $\partial/\partial x$, $\partial/\partial y$, and $\partial/\partial z$ commute with each other, we obtain

$$\left[-\frac{\hbar^2}{2m}\nabla^2, \hat{L}_x\right] = 0. \tag{16.56}$$

Similarly, we can show that

$$\left[-\frac{\hbar^2}{2m}\nabla^2, \hat{L}_y\right] = \left[-\frac{\hbar^2}{2m}\nabla^2, \hat{L}_z\right] = 0. \tag{16.57}$$

Consider now the commutator involving the potential energy

$$\left[\hat{V}(r), \hat{L}_x\right] = -i\hbar\left[\hat{V}(r), \left(y\frac{\partial}{\partial z} - z\frac{\partial}{\partial y}\right)\right]$$

$$= -i\hbar\left[\hat{V}(r)\left(y\frac{\partial}{\partial z} - z\frac{\partial}{\partial y}\right) - \left(y\frac{\partial}{\partial z} - z\frac{\partial}{\partial y}\right)\hat{V}(r)\right]$$

$$= -i\hbar\left[Vy\frac{\partial}{\partial z} - Vz\frac{\partial}{\partial y} - y\frac{\partial V}{\partial z} + z\frac{\partial V}{\partial y} - yV\frac{\partial}{\partial z} + zV\frac{\partial}{\partial y}\right]$$

$$= -i\hbar\left(z\frac{\partial V}{\partial y} - y\frac{\partial V}{\partial z}\right). \tag{16.58}$$

Since $\hat{V}(r)$ depends only on r, we have

$$\left(z\frac{\partial V}{\partial y} - y\frac{\partial V}{\partial z}\right) = z\frac{\partial V}{\partial r}\frac{\partial r}{\partial y} - y\frac{\partial V}{\partial r}\frac{\partial r}{\partial z}. \qquad (16.59)$$

If r is the position of an arbitrary point in the x, y, z coordinates, we have

$$r = \sqrt{x^2 + y^2 + z^2}, \qquad (16.60)$$

and then

$$\frac{\partial r}{\partial y} = \frac{y}{r}, \qquad \frac{\partial r}{\partial z} = \frac{z}{r}. \qquad (16.61)$$

Hence,

$$z\frac{\partial V}{\partial r}\frac{\partial r}{\partial y} - y\frac{\partial V}{\partial r}\frac{\partial r}{\partial z} = \frac{\partial V}{\partial r}\left(z\frac{y}{r} - y\frac{z}{r}\right) = 0. \qquad (16.62)$$

Thus,

$$\left[\hat{V}(r), \hat{L}_x\right] = 0. \qquad (16.63)$$

Similarly, we can show that

$$\left[\hat{V}(r), \hat{L}_y\right] = \left[\hat{V}(r), \hat{L}_y\right] = 0. \qquad (16.64)$$

In summary, since

$$\left[-\frac{\hbar^2}{2m}\nabla^2, \hat{L}_x\right] = \left[-\frac{\hbar^2}{2m}\nabla^2, \hat{L}_y\right] = \left[-\frac{\hbar^2}{2m}\nabla^2, \hat{L}_z\right] = 0,$$

$$\left[\hat{V}(r), \hat{L}_x\right] = \left[\hat{V}(r), \hat{L}_y\right] = \left[\hat{V}(r), \hat{L}_z\right] = 0, \qquad (16.65)$$

we have

$$\left[\hat{H}, \hat{\vec{L}}\right] = 0. \qquad (16.66)$$

Problem 16.6

Transition dipole moments

The electron in a hydrogen atom can be in two states of the form

$$\Psi_1(r) = \sqrt{2}Ne^{-r/a_o},$$

$$\Psi_2(r) = \frac{N}{4a_o}re^{-r/(2a_o)}\cos\theta, \qquad (16.67)$$

where $r = (x^2 + y^2 + z^2)^{\frac{1}{2}}$, $\cos \theta = z/r$, $N = 1/\sqrt{2\pi a_o^3}$, and a_o is the Bohr radius. Using the spherical coordinates, in which

$$x = r \sin \theta \cos \phi,$$
$$y = r \sin \theta \sin \phi,$$
$$z = r \cos \theta, \tag{16.68}$$

and

$$\int dV = \int_0^\infty \int_0^\pi \int_0^{2\pi} r^2 \sin \theta \, dr d\theta d\phi, \tag{16.69}$$

(a) show that the functions $\Psi_1(r)$, $\Psi_2(r)$ are orthogonal.
(b) Calculate the matrix element $(\Psi_1(r), \hat{r}\Psi_2(r))$ of the position operator $\hat{r}$ between the states $\Psi_1(r)$ and $\Psi_2(r)$.
    (The matrix element is related to the atomic electric dipole moment between the states $\Psi_1(r)$ and $\Psi_2(r)$, defined as $(\Psi_1(r), \hat{\mu}\Psi_2(r)) = e(\Psi_1(r), \hat{r}\Psi_2(r))$.)
(c) Show that the average values of the kinetic and potential energies in the state $\Psi_1(r)$ satisfy the relation $\langle E_k \rangle = -\frac{1}{2}\langle V \rangle$.

Solution (a)

Two functions are orthogonal when the scalar product

$$(\Psi_1(r), \Psi_2(r)) = \int \Psi_1^*(r)\Psi_2(r)dV = 0. \tag{16.70}$$

Calculate the integral

$$\int \Psi_1^*(r)\Psi_2(r)dV = \frac{\sqrt{2}N^2}{4a_o} \int r \cos \theta \, e^{-3r/2a_o} dV$$

$$= \frac{\sqrt{2}N^2}{4a_o} \int_0^\infty \int_0^\pi \int_0^{2\pi} r^3 \sin \theta \cos \theta \, e^{-3r/2a_o} dr d\theta d\phi$$

$$= \frac{2\pi \sqrt{2}N^2}{4a_o} \int_0^\infty \int_0^\pi r^3 \sin \theta \cos \theta \, e^{-3r/2a_o} dr d\theta.$$
$$\tag{16.71}$$

Consider the integral over θ:

$$\int_0^\pi \sin \theta \cos \theta \, d\theta. \tag{16.72}$$

Let $\sin \theta = x$. Then, $\cos \theta d\theta = dx$ and the integral takes the form

$$\int_0^\pi \sin \theta \cos \theta \, d\theta = \int_0^0 x dx = 0. \qquad (16.73)$$

Thus, the functions $\Psi_1(r)$, $\Psi_2(r)$ are orthogonal.

Solution (b)

From the definition of the matrix element, we have

$$\left(\Psi_1(r), \hat{\vec{r}} \, \Psi_2(r) \right) = \int \Psi_1^*(r) \hat{\vec{r}} \, \Psi_2(r) dV$$

$$= \vec{i} \int \Psi_1^*(r) x \Psi_2(r) dV + \vec{j} \int \Psi_1^*(r) y \Psi_2(r) dV$$

$$+ \vec{k} \int \Psi_1^*(r) z \Psi_2(r) dV. \qquad (16.74)$$

In spherical coordinates, and substituting the explicit forms of the functions $\Psi_1(r)$, $\Psi_2(r)$, the integrals take the form

$$\left(\Psi_1(r), \hat{\vec{r}} \, \Psi_2(r) \right)$$

$$= \vec{i} \frac{\sqrt{2}N^2}{4a_o} \int_0^\infty \int_0^\pi \int_0^{2\pi} r^4 \sin^2 \theta \cos \theta \cos \phi \, e^{-3r/2a_o} dr d\theta d\phi$$

$$+ \vec{j} \frac{\sqrt{2}N^2}{4a_o} \int_0^\infty \int_0^\pi \int_0^{2\pi} r^4 \sin^2 \theta \cos \theta \sin \phi \, e^{-3r/2a_o} dr d\theta d\phi$$

$$+ \vec{k} \frac{\sqrt{2}N^2}{4a_o} \int_0^\infty \int_0^\pi \int_0^{2\pi} r^4 \sin\theta \cos^2\theta \, e^{-3r/2a_o} dr d\theta d\phi. \quad (16.75)$$

Since

$$\int_0^{2\pi} \sin \phi d\phi = \int_0^{2\pi} \cos \phi d\phi = 0, \qquad (16.76)$$

the x and y components of the matrix element are zero. Hence

$$\left(\Psi_1(r), \hat{\vec{r}} \, \Psi_2(r) \right) = \vec{k} \frac{\sqrt{2}N^2}{4a_o} \int_0^\infty \int_0^\pi \int_0^{2\pi} r^4 \sin \theta \cos^2 \theta \, e^{-3r/2a_o} dr d\theta d\phi$$

$$= \vec{k} \frac{2\pi \sqrt{2}N^2}{4a_o} \int_0^\infty \int_0^\pi r^4 \sin \theta \cos^2 \theta \, e^{-3r/2a_o} dr d\theta$$

$$= \vec{k} \frac{2\pi \sqrt{2}N^2}{4a_o} \int_0^\infty r^4 e^{-3r/2a_o} dr \int_0^\pi \sin \theta \cos^2 \theta d\theta.$$

$$(16.77)$$

We will calculate separately the integrals over r and θ.

Let $\cos\theta = x$. Then, $-\sin\theta d\theta = dx$, and the integral over θ gives

$$\int_0^\pi \cos^2\theta \sin\theta d\theta = \int_{-1}^1 x^2 dx = \frac{1}{3}x^3 \Big|_{-1}^1 = \frac{2}{3}. \qquad (16.78)$$

Thus,

$$\left(\Psi_1(r), \hat{\vec{r}}\,\Psi_2(r)\right) = \vec{k}\frac{\pi\sqrt{2}N^2}{3a_o}\int_0^\infty r^4 e^{-3r/2a_o}dr. \qquad (16.79)$$

The remaining integral over r is easily evaluated, e.g., by parts, and gives

$$\int_0^\infty r^4 e^{-\beta r}dr = \frac{4!}{\beta^5}, \qquad (16.80)$$

where $\beta = 3/2a_o$. Hence, substituting for $N = 1/\sqrt{2\pi a_o^3}$, we get

$$\left(\Psi_1(r), \hat{\vec{r}}\,\Psi_2(r)\right) = \vec{k}\frac{\pi\sqrt{2}N^2}{3a_o}\frac{4!}{\beta^5} = \vec{k}\frac{\pi\sqrt{2}N^2}{3a_o}\frac{24a_o^5 \times 32}{243}$$

$$= \frac{128\sqrt{2}a_o}{243}\vec{k}. \qquad (16.81)$$

Solution (c)

The function $\Psi_1(r)$ can be written as

$$\Psi_1(r) = Ae^{-\alpha r}, \qquad (16.82)$$

where $A = \sqrt{2}N$ and $\alpha = 1/a_o$.

Consider the kinetic energy

$$\hat{E}_k = \frac{1}{2m}\hat{p}^2 = -\frac{\hbar^2}{2m}\nabla^2, \qquad (16.83)$$

where

$$\nabla^2 \equiv \frac{\partial^2}{\partial x^2} + \frac{\partial^2}{\partial y^2} + \frac{\partial^2}{\partial z^2}. \qquad (16.84)$$

Hence, the average kinetic energy in the state $\Psi_1(r)$ is

$$\langle\hat{E}_k\rangle = \int \Psi_1^*(r)\hat{E}_k\Psi_1(r)dV = -\frac{\hbar^2}{2m}A^2\int dV\, e^{-\alpha r}\nabla^2 e^{-\alpha r}$$

$$= -\frac{\hbar^2}{2m}A^2\int_0^\infty\int_0^\pi\int_0^{2\pi} dr d\theta d\phi\, r^2\sin\theta\, e^{-\alpha r}\nabla^2 e^{-\alpha r}$$

$$= -\frac{4\pi\hbar^2}{2m}A^2\int_0^\infty dr\, r^2 e^{-\alpha r}\nabla^2 e^{-\alpha r}. \qquad (16.85)$$

First, calculate $\nabla^2 e^{-\alpha r}$:

$$\frac{\partial}{\partial x}e^{-\alpha r} = -\frac{\alpha x}{r}e^{-\alpha r},$$

$$\frac{\partial^2}{\partial x^2}e^{-\alpha r} = -\alpha\frac{\partial}{\partial x}\left(\frac{x}{r}e^{-\alpha r}\right) = -\alpha\left(\frac{y^2+z^2}{r^3} - \frac{\alpha x^2}{r^2}\right)e^{-\alpha r}.$$

$$(16.86)$$

Similarly,

$$\frac{\partial^2}{\partial y^2}e^{-\alpha r} = -\alpha\left(\frac{x^2+z^2}{r^3} - \frac{\alpha y^2}{r^2}\right)e^{-\alpha r},$$

$$\frac{\partial^2}{\partial z^2}e^{-\alpha r} = -\alpha\left(\frac{x^2+y^2}{r^3} - \frac{\alpha z^2}{r^2}\right)e^{-\alpha r}.$$

$$(16.87)$$

Thus,

$$\nabla^2 e^{-\alpha r} = \frac{\partial^2}{\partial x^2}e^{-\alpha r} + \frac{\partial^2}{\partial y^2}e^{-\alpha r} + \frac{\partial^2}{\partial z^2}e^{-\alpha r}$$

$$= -\frac{2\alpha}{r}e^{-\alpha r} + \alpha^2 e^{-\alpha r}.$$

$$(16.88)$$

Hence, the average kinetic energy is

$$\langle \hat{E}_k \rangle = -\frac{2\pi\hbar^2}{m}A^2\left[-2\alpha\int_0^\infty dr\, r e^{-2\alpha r} + \alpha^2\int_0^\infty dr r^2 e^{-2\alpha r}\right]$$

$$= -\frac{2\pi\hbar^2}{m}A^2\left[-2\alpha\frac{1}{4\alpha^2} + \alpha^2\frac{2}{8\alpha^3}\right] = \frac{\hbar^2 A^2\pi}{2m\alpha}.$$

$$(16.89)$$

Since $A^2 = 1/(\pi a_o^3)$ and $\alpha = 1/a_o$, we get

$$\langle \hat{E}_k \rangle = \frac{\hbar^2 A^2\pi}{2m\alpha} = \frac{\hbar^2\pi}{2m}\frac{1}{\pi a_o^3}a_o = \frac{\hbar^2}{2m a_o^2}.$$

$$(16.90)$$

Consider now the potential energy defined as

$$\hat{V} = -\frac{e^2}{4\pi\varepsilon_0}\frac{1}{r} = -\frac{\eta}{r},$$

$$(16.91)$$

where $\eta = e^2/(4\pi\varepsilon_0)$.

The average potential energy in the state $\Psi_1(r)$ is given by

$$\langle \hat{V} \rangle = \int \Psi_1^*(r)\hat{V}\Psi_1(r)dV = -\eta A^2\int dV\, e^{-\alpha r}\frac{1}{r}e^{-\alpha r}$$

$$= -4\pi\eta A^2\int_0^\infty dr\, r e^{-2\alpha r} = -4\pi\eta A^2\frac{1}{4\alpha^2}$$

$$= -\pi\eta\frac{1}{\pi a_o^3}a_o^2 = -\frac{\eta}{a_o}.$$

$$(16.92)$$

Since

$$\eta = \frac{e^2}{4\pi\varepsilon_0} \quad \text{and} \quad a_0 = \frac{4\pi\varepsilon_0\hbar^2}{me^2}, \qquad (16.93)$$

we finally obtain

$$\langle\hat{V}\rangle = -\frac{\eta}{a_0} = -\frac{e^2}{4\pi\varepsilon_0}\frac{1}{a_0^2}a_0 = -\frac{e^2}{4\pi\varepsilon_0}\frac{1}{a_0^2}\frac{4\pi\varepsilon_0\hbar^2}{me^2} = -\frac{\hbar^2}{ma_0^2}.$$
$$(16.94)$$

We have found $\langle\hat{V}\rangle$ calculating the average value from the definition of the quantum expectation (average) value. However, there is a much quicker way to find $\langle\hat{V}\rangle$, simply by using the relation

$$\langle\hat{V}\rangle = \langle E\rangle - \langle\hat{E}_k\rangle = E - \langle\hat{E}_k\rangle, \qquad (16.95)$$

where

$$E = -\frac{1}{4\pi\varepsilon_0}\frac{e^2}{2a_0} = -\frac{1}{4\pi\varepsilon_0}\frac{e^2}{2a_0^2}a_0 = -\frac{1}{4\pi\varepsilon_0}\frac{e^2}{2a_0^2}\frac{4\pi\varepsilon_0\hbar^2}{me^2} = -\frac{\hbar^2}{2ma_0^2}$$
$$(16.96)$$

is the total energy of the electron in the state $\Psi_1(r)$. Thus,

$$\langle\hat{V}\rangle = E - \langle\hat{E}_k\rangle = -\frac{\hbar^2}{2ma_0^2} - \frac{\hbar^2}{2ma_0^2} = -\frac{\hbar^2}{ma_0^2}. \qquad (16.97)$$

Hence,

$$\langle\hat{V}\rangle = -2\langle\hat{E}_k\rangle \quad \text{i.e.,} \quad \langle\hat{E}_k\rangle = -\frac{1}{2}\langle\hat{V}\rangle. \qquad (16.98)$$

Problem 16.7

The wave functions of the electron in the states $n = 1$ and $n = 2$, $l = 1, m = 0$ of the hydrogen atom are

$$\Psi_{100} = \frac{1}{\sqrt{\pi a_0^3}}e^{-r/a_0},$$

$$\Psi_{210} = \frac{1}{\sqrt{32\pi a_0^3}}\frac{r}{a_0}e^{-r/(2a_0)}\cos\theta, \qquad (16.99)$$

where a_0 is the Bohr radius.

(a) Calculate the standard deviation $\sigma^2 = \langle r^2\rangle - \langle r\rangle^2$ of the position of the electron in these two states to determine in which of these states, the electron is more stable in the position.

(b) The electron is found in the state

$$\Psi = \sqrt{\frac{8}{\pi a_o^3}}\, e^{-2r/a_o}. \tag{16.100}$$

Determine what is the probability that the state Ψ is the ground state $(n = 1)$ of the hydrogen atom.

Solution (a)

Calculate first the standard deviation in the state Ψ_{100}. From the definition of expectation value, we find

$$\langle r \rangle = \int \Psi_{100}^* r \Psi_{100} dV = \frac{4\pi}{\pi a_o^3} \int_0^\infty dr\, r^3 e^{-2r/a_o} = \frac{4}{a_o^3}\frac{3a_o^4}{8} = \frac{3}{2}a_o,$$

$$\langle r^2 \rangle = \int \Psi_{100}^* r^2 \Psi_{100} dV = \frac{4\pi}{\pi a_o^3} \int_0^\infty dr\, r^4 e^{-2r/a_o} = \frac{4}{a_o^3}\frac{3a_o^5}{4} = 3a_o^2.$$

$$\tag{16.101}$$

Thus, the standard deviation in the state Ψ_{100} is

$$\sigma_{100}^2 = \langle r^2 \rangle - \langle r \rangle^2 = 3a_o^2 - \frac{9}{4}a_o^2 = \frac{3}{4}a_o^2. \tag{16.102}$$

In the case of the state Ψ_{210}, the average values $\langle r \rangle$ and $\langle r^2 \rangle$ are given by the following double integrals:

$$\langle r \rangle = \int \Psi_{210}^* r \Psi_{210} dV = 2\pi \int_0^\infty dr \int_0^\pi d\theta \sin\theta\, \Psi_{210}^* r^3 \Psi_{210},$$

$$\langle r^2 \rangle = \int \Psi_{210}^* r^2 \Psi_{210} dV = 2\pi \int_0^\infty dr \int_0^\pi d\theta \sin\theta\, \Psi_{210}^* r^4 \Psi_{210}.$$

$$\tag{16.103}$$

Substituting the explicit form of Ψ_{210}, we get for $\langle r \rangle$

$$\langle r \rangle = 2\pi \int_0^\infty dr \int_0^\pi d\theta \sin\theta\, \Psi_{210}^* r^3 \Psi_{210}$$

$$= \frac{2\pi}{32\pi a_o^5} \int_0^\infty dr \int_0^\pi d\theta \cos^2\theta \sin\theta\, r^5 e^{-r/a_o}$$

$$= \frac{1}{16a_o^5} \int_0^\infty dr\, r^5 e^{-r/a_o} \int_0^\pi d\theta \cos^2\theta \sin\theta. \tag{16.104}$$

Since

$$\int_0^\pi d\theta \cos^2\theta \sin\theta = \int_{-1}^1 x^2 dx = \frac{2}{3}, \qquad (16.105)$$

and

$$\int_0^\infty dr r^5 e^{-r/a_0} = 120 a_o^6, \qquad (16.106)$$

we have

$$\langle r \rangle = \frac{1}{16 a_o^5}\frac{2}{3}120 a_o^6 = 5a_o. \qquad (16.107)$$

Similarly, for $\langle r^2 \rangle$, we get

$$
\begin{aligned}
\langle r^2 \rangle &= 2\pi \int_0^\infty dr \int_0^\pi d\theta \sin\theta \, \Psi_{210}^* r^4 \Psi_{210} \\
&= \frac{2\pi}{32\pi a_o^5} \int_0^\infty dr \int_0^\pi d\theta \cos^2\theta \sin\theta \, r^6 e^{-r/a_0} \\
&= \frac{1}{16 a_o^5} \int_0^\infty dr r^6 e^{-r/a_0} \int_0^\pi d\theta \cos^2\theta \sin\theta \\
&= \frac{1}{16 a_o^5}\frac{2}{3} \int_0^\infty dr r^6 e^{-r/a_0} = \frac{1}{24 a_o^5}720 a_o^7 = 30 a_o^2.
\end{aligned}
$$
$$\qquad (16.108)$$

Hence, the standard deviation in the state Ψ_{210} is

$$\sigma_{210}^2 = 30 a_o^2 - 25 a_o^2 = 5a_o^2. \qquad (16.109)$$

Since $\sigma_{100}^2 < \sigma_{210}^2$, the electron is more stable in the state Ψ_{100} than in the state Ψ_{210}.

Solution (b)

The probability is determined by the scalar product of the state Ψ and the state Ψ_{100}. In other words, the probability tells us to what extent the state Ψ overlaps with the state Ψ_{100}. In the modern terminology, it is called fidelity.

From the definition of the scalar product of two states, we have in the spherical coordinates

$$(\Psi, \Psi_{100}) = 4\pi \int_0^\infty dr r^2 \Psi^* \Psi_{100} = \frac{8\sqrt{2}\pi}{\pi a_o^3} \int_0^\infty dr\, r^2 e^{-3r/a_o}$$

$$= \frac{8\sqrt{2}}{a_o^3} \frac{2a_o^3}{27} = \frac{16\sqrt{2}}{27} \approx 0.84. \qquad\qquad (16.110)$$

Thus, with probability $P = 0.84$, the state Ψ can be considered the ground state of the hydrogen atom.

Chapter 17

Quantum Theory of Two Coupled Particles

Problem 17.1

Suppose that a particle of mass m can rotate around a fixed point A, such that $r = $ constant and $\theta = \pi/2 = $ constant.

(a) Show that the motion of the particle is quantized.
(b) Show that the only acceptable solutions to the wave function of the particle are those corresponding to positive energies $(E > 0)$ of the particle.

Solution (a)

Consider the rotation in spherical coordinates. Since r and θ are constant, the rotation depends only on the azimuthal angle ϕ. In this case, the Schrödinger equation simplifies to

$$-\frac{\hbar^2}{2mr^2}\frac{\partial^2 \Psi}{\partial \phi^2} = E\Psi, \tag{17.1}$$

Problems and Solutions in Quantum Physics
Zbigniew Ficek
Copyright © 2016 Pan Stanford Publishing Pte. Ltd.
ISBN 978-981-4669-36-8 (Hardcover), 978-981-4669-37-5 (eBook)
www.panstanford.com

which can be written as

$$\frac{\partial^2 \Psi}{\partial \phi^2} = -\frac{2mr^2 E}{\hbar^2} \Psi. \tag{17.2}$$

We see that the wave function of the rotating mass satisfies a simple differential equation of harmonic motion

$$\frac{\partial^2 \Psi}{\partial \phi^2} = -\beta^2 \Psi, \tag{17.3}$$

whose solution is

$$\Psi(\phi) = A e^{i\beta\phi}, \tag{17.4}$$

where A is a constant and $\beta^2 = \frac{2mr^2 E}{\hbar^2}$.

Since in rotation $\Psi(\phi) = \Psi(\phi + 2\pi)$, we find that this is satisfied when

$$e^{2\pi i\beta} = 1, \tag{17.5}$$

i.e., when β is an integer ($\beta = 0, \pm 1, \pm 2, \ldots$).

Hence, β^2 is not an arbitrary number but an integer. This shows that the energy in the rotation is quantized.

Solution (b)

For $E < 0$, the Schrödinger equation takes the form

$$\frac{\partial^2 \Psi}{\partial \phi^2} = \frac{2mr^2 |E|}{\hbar^2} \Psi = \beta^2 \Psi \qquad \beta^2 > 0. \tag{17.6}$$

The solution to the above differential equation is of the form

$$\Psi(\phi) = A e^{\beta\phi} + B e^{-\beta\phi}. \tag{17.7}$$

This is a damped function that does not describe rotation. Thus, it is not an acceptable solution to the wave function of the rotating mass.

For $E > 0$, the Schrödinger equation is of the form

$$\frac{\partial^2 \Psi}{\partial \phi^2} = -\frac{2mr^2 E}{\hbar^2} \Psi = -\beta^2 \Psi \qquad \beta^2 > 0. \tag{17.8}$$

The solution to the above differential equation is of the form

$$\Psi(\phi) = A e^{i\beta\phi}, \tag{17.9}$$

which describes rotation. Thus, it is an acceptable solution to the wave function of the rotating mass.

Chapter 18

Time-Independent Perturbation Theory

Problem 18.1

In an orthonormal basis, a linear operator $\hat{A}$ is represented by the matrix

$$\hat{A} = \begin{pmatrix} 2\lambda & 1+\lambda \\ 1+\lambda & \lambda \end{pmatrix}, \tag{18.1}$$

where λ is a small real parameter ($\lambda \ll 1$). The operator $\hat{A}$ can be written as the sum of two operators, $\hat{A} = \hat{A}_0 + \lambda \hat{V}$, where

$$\hat{A}_0 = \begin{pmatrix} 0 & 1 \\ 1 & 0 \end{pmatrix}, \quad \hat{V} = \begin{pmatrix} 2 & 1 \\ 1 & 1 \end{pmatrix}. \tag{18.2}$$

Using the first-order perturbation theory, find the eigenvalues and eigenvectors of $\hat{A}$ in terms of the eigenvalues and eigenvectors of $\hat{A}_0$.

Notice that $\hat{A}_0$ is of the same form as the x-component of the electron spin, $\hat{\sigma}_x$.

Solution

The unperturbed states are the eigenstates of the operator $\hat{A}_0$. Since $\hat{A}_0 = \hat{\sigma}_x$, the unperturbed eigenstates and the corresponding

Problems and Solutions in Quantum Physics
Zbigniew Ficek
Copyright © 2016 Pan Stanford Publishing Pte. Ltd.
ISBN 978-981-4669-36-8 (Hardcover), 978-981-4669-37-5 (eBook)
www.panstanford.com

eigenvalues are (for details see Tutorial Problem 13.4)

$$|\phi_1^{(0)}\rangle = \frac{1}{\sqrt{2}} \begin{pmatrix} 1 \\ 1 \end{pmatrix}, \quad E_1^{(0)} = 1.$$

$$|\phi_2^{(0)}\rangle = \frac{1}{\sqrt{2}} \begin{pmatrix} 1 \\ -1 \end{pmatrix}, \quad E_2^{(0)} = -1. \tag{18.3}$$

The first-order correction to the eigenvalue $E_1^{(0)}$ is equal to the expectation value of $\hat{V}$ in the state $|\phi_1^{(0)}\rangle$. Hence,

$$E_1^{(1)} = \langle \phi_1^{(0)}|\hat{V}|\phi_1^{(0)}\rangle = \frac{1}{2} (1\ 1) \begin{pmatrix} 2 & 1 \\ 1 & 1 \end{pmatrix} \begin{pmatrix} 1 \\ 1 \end{pmatrix}$$

$$= \frac{1}{2} (1\ 1) \begin{pmatrix} 3 \\ 2 \end{pmatrix} = \frac{1}{2} (3 + 2) = \frac{5}{2}. \tag{18.4}$$

Similarly, the first-order correction to the eigenvalue $E_2^{(0)}$ is

$$E_2^{(1)} = \langle \phi_2^{(0)}|\hat{V}|\phi_2^{(0)}\rangle = \frac{1}{2} (1\ -1) \begin{pmatrix} 2 & 1 \\ 1 & 1 \end{pmatrix} \begin{pmatrix} 1 \\ -1 \end{pmatrix}$$

$$= \frac{1}{2} (1\ -1) \begin{pmatrix} 1 \\ 0 \end{pmatrix} = \frac{1}{2}. \tag{18.5}$$

Thus, the eigenvalues of $\hat{A}$ to the first order in λ are

$$E_1 = E_1^{(0)} + \lambda E_1^{(1)} = 1 + \frac{5}{2}\lambda,$$

$$E_2 = E_2^{(0)} + \lambda E_2^{(1)} = -1 + \frac{1}{2}\lambda. \tag{18.6}$$

Calculate now the first-order corrections to the eigenvectors. The first-order correction to the eigenvector $|\phi_1^{(0)}\rangle$ is

$$|\phi_1^{(1)}\rangle = \frac{\langle \phi_2^{(0)}|\hat{V}|\phi_1^{(0)}\rangle}{E_1^{(0)} - E_2^{(0)}} |\phi_2^{(0)}\rangle = \frac{\frac{1}{2}(1\ -1) \begin{pmatrix} 2 & 1 \\ 1 & 1 \end{pmatrix} \begin{pmatrix} 1 \\ 1 \end{pmatrix}}{1 - (-1)} |\phi_2^{(0)}\rangle$$

$$= \frac{\frac{1}{2}(1\ -1) \begin{pmatrix} 3 \\ 2 \end{pmatrix}}{2} |\phi_2^{(0)}\rangle = \frac{1}{4} (3 - 2) |\phi_2^{(0)}\rangle = \frac{1}{4}|\phi_2^{(0)}\rangle. \tag{18.7}$$

Similarly, the first-order correction to the eigenvector $|\phi_2^{(0)}\rangle$ is

$$
|\phi_2^{(1)}\rangle = \frac{\langle \phi_1^{(0)}|\hat{V}|\phi_2^{(0)}\rangle}{E_2^{(0)} - E_1^{(0)}}|\phi_1^{(0)}\rangle = \frac{\frac{1}{2}(1 \ 1)\begin{pmatrix} 2 & 1 \\ 1 & 1 \end{pmatrix}\begin{pmatrix} 1 \\ -1 \end{pmatrix}}{(-1) - 1}|\phi_1^{(0)}\rangle
$$

$$
= \frac{\frac{1}{2}(1 \ 1)\begin{pmatrix} 1 \\ 0 \end{pmatrix}}{-2}|\phi_1^{(0)}\rangle = -\frac{1}{4}|\phi_1^{(0)}\rangle. \tag{18.8}
$$

Thus, the eigenvectors of $\hat{A}$ to the first order in λ are

$$
|\phi_1\rangle = |\phi_1^{(0)}\rangle + \lambda|\phi_1^{(1)}\rangle = |\phi_1^{(0)}\rangle + \frac{1}{4}\lambda|\phi_2^{(0)}\rangle,
$$

$$
|\phi_2\rangle = |\phi_2^{(0)}\rangle + \lambda|\phi_2^{(1)}\rangle = |\phi_2^{(0)}\rangle - \frac{1}{4}\lambda|\phi_1^{(0)}\rangle. \tag{18.9}
$$

Chapter 19

Time-Dependent Perturbation Theory

Problem 19.1

Consider a two-level atom represented by the spin operators $\hat{\sigma}^{\pm}$, $\hat{\sigma}_z$, interacting with a one-dimensional harmonic oscillator, represented by the creation and annihilation operators $\hat{a}^{\dagger}$ and $\hat{a}$. The Hamiltonian of the system is given by

$$\hat{H} = \frac{1}{2}\hbar\omega_0\hat{\sigma}_z + \hbar\omega_0\left(\hat{a}^{\dagger}\hat{a} + \frac{1}{2}\right) - \frac{1}{2}i\hbar g\left(\hat{\sigma}^{+}\hat{a} - \hat{\sigma}^{-}\hat{a}^{\dagger}\right). \quad (19.1)$$

The Hamiltonian can be written as

$$\hat{H} = \hat{H}_0 + \hat{V}, \quad (19.2)$$

where

$$\hat{H}_0 = \frac{1}{2}\hbar\omega_0\hat{\sigma}_z + \hbar\omega_0\left(\hat{a}^{\dagger}\hat{a} + \frac{1}{2}\right),$$

$$\hat{V} = -\frac{1}{2}i\hbar g\left(\hat{\sigma}^{+}\hat{a} - \hat{\sigma}^{-}\hat{a}^{\dagger}\right). \quad (19.3)$$

The eigenstates of $\hat{H}_0$ are product states

$$|\phi_n\rangle = |n\rangle|1\rangle, \quad |\phi_{n-1}\rangle = |n-1\rangle|2\rangle, \quad (19.4)$$

where $|n\rangle$ is the photon number state of the harmonic oscillator and $|1\rangle$, $|2\rangle$ are the energy states of the atom.

Problems and Solutions in Quantum Physics
Zbigniew Ficek
Copyright © 2016 Pan Stanford Publishing Pte. Ltd.
ISBN 978-981-4669-36-8 (Hardcover), 978-981-4669-37-5 (eBook)
www.panstanford.com

(a) Write the state vector of the system in terms of the eigenstates of $\hat{H}_0$.

(b) Assume that initially at $t = 0$, the system was in the state $|\phi_n\rangle$. Find the probability, using the time-dependent perturbation theory, that after a time t, the system can be found in the state $|\phi_{n-1}\rangle$.

Solution (a)

The state vector of the system is given by

$$|\Psi(t)\rangle = \sum_m c_m(t)e^{-\frac{i}{\hbar}E_m t}|\phi_m\rangle, \quad m = n, n-1, \quad (19.5)$$

where E_m is the energy of the state $|\phi_m\rangle$. The energy of the state $|\phi_n\rangle$ is

$$E_n = \langle\phi_n|\hat{H}_0|\phi_n\rangle = \frac{1}{2}\hbar\omega_0\langle 1|\hat{\sigma}_z|1\rangle + \hbar\omega_0\langle n|\left(\hat{a}^\dagger\hat{a} + \frac{1}{2}\right)|n\rangle$$

$$= -\frac{1}{2}\hbar\omega_0 + n\hbar\omega_0 + \frac{1}{2}\hbar\omega_0 = n\hbar\omega_0. \quad (19.6)$$

The energy of the state $|\phi_{n-1}\rangle$ is

$$E_{n-1} = \langle\phi_{n-1}|\hat{H}_0|\phi_{n-1}\rangle$$

$$= \frac{1}{2}\hbar\omega_0\langle 2|\hat{\sigma}_z|2\rangle + \hbar\omega_0\langle n-1|\left(\hat{a}^\dagger\hat{a} + \frac{1}{2}\right)|n-1\rangle$$

$$= \frac{1}{2}\hbar\omega_0 + (n-1)\hbar\omega_0 + \frac{1}{2}\hbar\omega_0 = n\hbar\omega_0. \quad (19.7)$$

Thus, $E_n = E_{n-1}$, i.e., the states $|\phi_n\rangle$ and $|\phi_{n-1}\rangle$ are degenerate. Hence, the state vector of the system is of the form

$$|\Psi(t)\rangle = \sum_m c_m(t)e^{-in\omega_0 t}|\phi_m\rangle, \quad m = n, n-1. \quad (19.8)$$

The unknown coefficients $c_m(t)$ can be determined using the time-dependent perturbation theory. We shall limit our calculations to the first-order corrections.

Since the interaction Hamiltonian $\hat{V}$ is independent of time, and assuming that $c_n^{(0)}(t) = c_n^{(0)}(0)$, the first-order corrections to the amplitudes $c_m(t)$ are

$$c_m^{(1)}(t) = -\frac{V_{mk}c_n^{(0)}(0)}{E_m - E_k}\left(e^{i\omega_{mk}t} - 1\right). \quad (19.9)$$

Hence, the first-order correction to $c_n(t)$ is

$$c_n^{(1)}(t) = -\frac{V_{n,n-1}c_{n-1}^{(0)}(0)}{E_n - E_{n-1}}\left(e^{i\omega_{n,n-1}t} - 1\right). \tag{19.10}$$

Since $E_n = E_{n-1} = n\hbar\omega_0$, we get using the Taylor expansion

$$c_n^{(1)}(t) = -\frac{V_{n,n-1}c_{n-1}^{(0)}(0)}{\hbar\omega_{n,n-1}}(1 + i\omega_{n,n-1}t + \ldots - 1) = -\frac{i}{\hbar}c_{n-1}^{(0)}(0)V_{n,n-1}t. \tag{19.11}$$

The explicit value of the matrix element $V_{n,n-1}$ is

$$
\begin{aligned}
V_{n,n-1} &= \langle\phi_n|\hat{V}|\phi_{n-1}\rangle \\
&= -\frac{1}{2}i\hbar g\langle 1|\langle n|\hat{\sigma}^+\hat{a}|2\rangle|n-1\rangle + \frac{1}{2}i\hbar g\langle 1|\langle n|\hat{\sigma}^-\hat{a}^\dagger|2\rangle|n-1\rangle \\
&= 0 + \frac{1}{2}i\hbar g\sqrt{n} = \frac{1}{2}i\hbar g\sqrt{n}, \tag{19.12}
\end{aligned}
$$

where we have used the results

$$\langle 1|\hat{\sigma}^+|2\rangle = 0, \quad \langle 1|\hat{\sigma}^-|2\rangle = 1, \quad \langle n|\hat{a}|n-1\rangle = 0, \quad \langle n|\hat{a}^\dagger|n-1\rangle = \sqrt{n}. \tag{19.13}$$

Thus,

$$c_n^{(1)}(t) = \frac{1}{2}c_{n-1}^{(0)}(0)g\sqrt{n}t. \tag{19.14}$$

Consider now the first-order correction to $c_{n-1}(t)$, which is given by

$$
\begin{aligned}
c_{n-1}^{(1)}(t) &= -\frac{V_{n-1,n}c_n^{(0)}(0)}{E_{n-1} - E_n}\left(e^{i\omega_{n-1,n}t} - 1\right) \\
&= -\frac{V_{n-1,n}c_n^{(0)}(0)}{\hbar\omega_{n-1,n}}(1 + i\omega_{n-1,n}t + \ldots - 1) \\
&= -\frac{i}{\hbar}c_n^{(0)}(0)V_{n-1,n}t. \tag{19.15}
\end{aligned}
$$

Calculating the value of the matrix element $V_{n-1,n}$, we get

$$
\begin{aligned}
V_{n-1,n} &= \langle\phi_{n-1}|\hat{V}|\phi_n\rangle \\
&= -\frac{1}{2}i\hbar g\langle 2|\langle n-1|\hat{\sigma}^+\hat{a}|1\rangle|n\rangle + \frac{1}{2}i\hbar g\langle 2|\langle n-1|\hat{\sigma}^-\hat{a}^\dagger|1\rangle|n\rangle \\
&= -\frac{1}{2}i\hbar g\sqrt{n} + 0 = -\frac{1}{2}i\hbar g\sqrt{n}. \tag{19.16}
\end{aligned}
$$

Thus,

$$c_{n-1}^{(1)}(t) = -\frac{1}{2}c_n^{(0)}(0)g\sqrt{n}t. \tag{19.17}$$

Hence, in the first order, the coefficients $c_n(t)$ and $c_{n-1}(t)$ are

$$c_n(t) = c_n(0) + \frac{1}{2}c_{n-1}(0)g\sqrt{n}t,$$

$$c_{n-1}(t) = c_{n-1}(0) - \frac{1}{2}c_n(0)g\sqrt{n}t. \qquad (19.18)$$

Thus, the state vector of the system, in the first order of the interaction, is of the form

$$|\Psi(t)\rangle = \left(c_n(0) + \frac{1}{2}c_{n-1}(0)g\sqrt{n}t\right)|\phi_n\rangle + \left(c_{n-1}(0) - \frac{1}{2}c_n(0)g\sqrt{n}t\right)|\phi_{n-1}\rangle. \qquad (19.19)$$

Solution (b)

The probability of a transition that after time t the system, initially at $t = 0$ in the state ϕ_n, will be found in the state ϕ_{n-1} is given by the absolute square of the amplitude $c_{n-1}^{(1)}(t)$:

$$P_{n \to n-1}(t) = |c_{n-1}^{(1)}(t)|^2 = \left|\frac{V_{n-1,n}}{E_{n-1} - E_n}\left(e^{i\omega_{n-1,n}t} - 1\right)\right|^2. \qquad (19.20)$$

Since

$$\left|e^{i\omega_{n-1,n}t} - 1\right|^2 = 4\sin^2\left(\frac{1}{2}\omega_{n-1,n}t\right), \qquad (19.21)$$

and

$$E_{n-1} - E_n = \hbar\omega_{n-1,n}, \qquad (19.22)$$

the transition probability simplifies to

$$P_{n \to n-1}(t) = \frac{|V_{n-1,n}|^2 t^2}{\hbar^2}\frac{\sin^2\left(\frac{1}{2}\omega_{n-1,n}t\right)}{\left(\frac{1}{2}\omega_{n-1,n}t\right)^2}. \qquad (19.23)$$

Since the states ϕ_n and ϕ_{n-1} are degenerate in energy, i.e., $\omega_{n-1,n} = 0$, the function $\sin^2 x/x^2 = 1$, and then

$$P_{n \to n-1}(t) = \frac{|V_{n-1,n}|^2 t^2}{\hbar^2} = \frac{1}{4}g^2 n t^2. \qquad (19.24)$$

The probability is proportional to the strength of the interaction, g^2, number of photons in the field, n, and the square of interaction time, t^2. Since $|V_{n-1,n}|^2 = |V_{n,n-1}|^2$, we see that the probability of transitions between the two states is the same in either direction.

Chapter 20

Relativistic Schrödinger Equation

Problem 20.1

Show that the Klein–Gordon equation for a free particle is invariant under the Lorentz transformation. The Lorentz transformation is given by

$$x' = \gamma(x - \beta ct),$$
$$y' = y,$$
$$z' = z,$$
$$ct' = \gamma(ct - \beta x), \qquad (20.1)$$

where $\gamma = \left(1 - \beta^2\right)^{-1/2}$ is the Lorentz factor, $\beta = u/c$, and u is the velocity an observed moves.

Solution

In order to show that the Klein–Gordon equation for a free particle is invariant under the Lorentz transformation, we have to demonstrate that the equation has the same form in both (t, x, y, z) and (t', x', y', z') coordinates.

Problems and Solutions in Quantum Physics
Zbigniew Ficek
Copyright © 2016 Pan Stanford Publishing Pte. Ltd.
ISBN 978-981-4669-36-8 (Hardcover), 978-981-4669-37-5 (eBook)
www.panstanford.com

Let us start from the Klein–Gordon equation in the (t, x, y, z) coordinates

$$\left(\square + \frac{m^2c^2}{\hbar^2}\right)\Psi = 0. \tag{20.2}$$

and using the Lorentz transformation (20.1), we shall demonstrate that in the (t', x', y', z') coordinates it has the form

$$\left(\square' + \frac{m^2c^2}{\hbar^2}\right)\Psi = 0. \tag{20.3}$$

We see that to demonstrate the invariance of the Klein–Gordon equation under the Lorentz transformation, it is enough to show that $\square\Psi = \square'\Psi$.

Since

$$\square\Psi = \frac{1}{c^2}\frac{\partial^2\Psi}{\partial t^2} - \nabla^2\Psi = \frac{\partial^2\Psi}{\partial(ct)^2} - \frac{\partial^2\Psi}{\partial x^2} - \frac{\partial^2\Psi}{\partial y^2} - \frac{\partial^2\Psi}{\partial z^2}, \tag{20.4}$$

we have to find how in the above equation, the second-order derivatives given in the (t, x, y, z) coordinates transform to those in the (t', x', y', z') coordinates.

Consider the first-order derivative over time. Since ct is a function of ct' and x', we apply the chain rule and obtain

$$\frac{\partial\Psi}{\partial(ct)} = \frac{\partial\Psi}{\partial(ct')}\frac{\partial(ct')}{\partial(ct)} + \frac{\partial\Psi}{\partial x'}\frac{\partial x'}{\partial(ct)}. \tag{20.5}$$

From Eq. (20.1), we have

$$\frac{\partial(ct')}{\partial(ct)} = \gamma, \quad \frac{\partial x'}{\partial(ct)} = -\beta\gamma. \tag{20.6}$$

Hence,

$$\frac{\partial\Psi}{\partial(ct)} = \gamma\frac{\partial\Psi}{\partial(ct')} - \beta\gamma\frac{\partial\Psi}{\partial x'}. \tag{20.7}$$

Then the second-order derivative over time is

$$\frac{\partial^2\Psi}{\partial(ct)^2} = \frac{\partial}{\partial(ct)}\frac{\partial\Psi}{\partial(ct)} = \left(\gamma\frac{\partial}{\partial(ct')} - \beta\gamma\frac{\partial}{\partial x'}\right)\left(\gamma\frac{\partial\Psi}{\partial(ct')} - \beta\gamma\frac{\partial\Psi}{\partial x'}\right)$$

$$= \gamma^2\left[\frac{\partial^2\Psi}{\partial(ct')^2} - \beta\frac{\partial^2\Psi}{\partial(ct')\partial x'} - \beta\frac{\partial^2\Psi}{\partial x'\partial(ct')} + \beta^2\frac{\partial^2\Psi}{\partial x'^2}\right]. \tag{20.8}$$

Consider now the first-order derivative over x:

$$\frac{\partial \Psi}{\partial x} = \frac{\partial \Psi}{\partial x'}\frac{\partial x'}{\partial x} + \frac{\partial \Psi}{\partial (ct')}\frac{\partial (ct')}{\partial x}. \tag{20.9}$$

Since

$$\frac{\partial (ct')}{\partial x} = -\beta\gamma, \quad \frac{\partial x'}{\partial x} = \gamma, \tag{20.10}$$

we get

$$\frac{\partial \Psi}{\partial x} = \gamma\frac{\partial \Psi}{\partial x'} - \beta\gamma\frac{\partial \Psi}{\partial (ct')}. \tag{20.11}$$

Then, the second-order derivative over x is

$$\frac{\partial^2 \Psi}{\partial x^2} = \frac{\partial}{\partial x}\frac{\partial \Psi}{\partial x} = \left(\gamma\frac{\partial}{\partial x'} - \beta\gamma\frac{\partial}{\partial (ct')}\right)\left(\gamma\frac{\partial \Psi}{\partial x'} - \beta\gamma\frac{\partial \Psi}{\partial (ct')}\right)$$

$$= \gamma^2\left[\frac{\partial^2 \Psi}{\partial x'^2} - \beta\frac{\partial^2 \Psi}{\partial x'\partial (ct')} - \beta\frac{\partial^2 \Psi}{\partial (ct')\partial x'} + \beta^2\frac{\partial^2 \Psi}{\partial (ct')^2}\right]. \tag{20.12}$$

Since $y' = y$ and $z' = z$, the second-order derivatives over y and z are

$$\frac{\partial^2 \Psi}{\partial y^2} = \frac{\partial^2 \Psi}{\partial y'^2}, \quad \frac{\partial^2 \Psi}{\partial z^2} = \frac{\partial^2 \Psi}{\partial z'^2}. \tag{20.13}$$

Collecting the results (20.8), (20.12), and (20.13), we get

$$\Box\Psi = \frac{\partial^2 \Psi}{\partial (ct)^2} - \frac{\partial^2 \Psi}{\partial x^2} - \frac{\partial^2 \Psi}{\partial y^2} - \frac{\partial^2 \Psi}{\partial z^2}$$

$$= \gamma^2\left[\frac{\partial^2 \Psi}{\partial (ct')^2} - \beta\frac{\partial^2 \Psi}{\partial (ct')\partial x'} - \beta\frac{\partial^2 \Psi}{\partial x'\partial (ct')} + \beta^2\frac{\partial^2 \Psi}{\partial x'^2}\right]$$

$$-\gamma^2\left[\frac{\partial^2 \Psi}{\partial x'^2} - \beta\frac{\partial^2 \Psi}{\partial x'\partial (ct')} - \beta\frac{\partial^2 \Psi}{\partial (ct')\partial x'} + \beta^2\frac{\partial^2 \Psi}{\partial (ct')^2}\right] - \frac{\partial^2 \Psi}{\partial y'^2}$$

$$-\frac{\partial^2 \Psi}{\partial z'^2} = \gamma^2(1 - \beta^2)\frac{\partial^2 \Psi}{\partial (ct')^2} - \gamma^2(1 - \beta^2)\frac{\partial^2 \Psi}{\partial x'^2} - \frac{\partial^2 \Psi}{\partial y'^2}$$

$$-\frac{\partial^2 \Psi}{\partial z'^2}. \tag{20.14}$$

However, $\gamma^2(1 - \beta^2) = 1$. Therefore

$$\Box\Psi = \frac{\partial^2 \Psi}{\partial (ct')^2} - \frac{\partial^2 \Psi}{\partial x'^2} - \frac{\partial^2 \Psi}{\partial y'^2} - \frac{\partial^2 \Psi}{\partial z'^2} = \Box'\Psi. \tag{20.15}$$

This shows that the Klein–Gordon equation has the same form in both coordinates. In other words, the Klein–Gordon equation is invariant under the Lorentz transformation.

Problem 20.2

Act on the Dirac equation

$$\left(E - c\vec{\alpha} \cdot \vec{p} - \beta mc^2\right) \Psi = 0 \tag{20.16}$$

with the operator

$$E + c\vec{\alpha} \cdot \vec{p} + \beta mc^2 \tag{20.17}$$

to find under which conditions the Dirac equation satisfies the relativistic energy relation

$$E^2 = c^2 p^2 + m^2 c^4. \tag{20.18}$$

Here, $\vec{\alpha} = \alpha_x \hat{i} + \alpha_y \hat{j} + \alpha_z \hat{k}$ is a three-dimensional Hermitian operator and β is a one-dimensional Hermitian operator. The operator β does not commute with any of the components of $\vec{\alpha}$.

Solution

This tutorial problem follows closely the derivation of the Dirac equation presented in the textbook. We particularly feel that the derivation should be discussed in details especially that the Dirac equation is not usually introduced at the basic level of quantum mechanics, but at the advanced level. We adopt here a simple vectorial formalism and show that the basic concepts of the relativistic Schrödinger equation can be easily understood in terms of the vector analysis and matrix multiplication.

Let us act on the equation

$$\left(E - c\vec{\alpha} \cdot \vec{p} - \beta mc^2\right) \Psi = 0 \tag{20.19}$$

from the left with the operator

$$E + c\vec{\alpha} \cdot \vec{p} + \beta mc^2. \tag{20.20}$$

We then have

$$\left(E - c\vec{\alpha} \cdot \vec{p} - \beta mc^2\right)\left(E + c\vec{\alpha} \cdot \vec{p} + \beta mc^2\right) \Psi = 0. \tag{20.21}$$

One might worry why the quantities $E - c\vec{\alpha} \cdot \vec{p} - \beta mc^2$ and $E + c\vec{\alpha} \cdot \vec{p} + \beta mc^2$ are called operators if they involve scalars and vectors only. The reason is that $\vec{\alpha}$ is a three-dimensional (vector) matrix and as such $\vec{\alpha}$ can be treated as the matrix representation of an operator $\hat{\alpha}$.

Performing the multiplication of the terms in Eq. (20.21), we get

$$\left\{ E^2 - c^2 \, (\vec{\alpha} \cdot \vec{p})^2 - mc^3 \, [(\vec{\alpha} \cdot \vec{p}) \beta + \beta \, (\vec{\alpha} \cdot \vec{p})] - \beta^2 m^2 c^4 \right\} \Psi = 0. \tag{20.22}$$

The scalar product $\vec{\alpha} \cdot \vec{p}$, appearing in the above expression, can be written in terms of components

$$\vec{\alpha} \cdot \vec{p} = \alpha_x \, p_x + \alpha_y \, p_y + \alpha_z \, p_z. \tag{20.23}$$

Squaring this expression gives

$$\begin{aligned}
(\vec{\alpha} \cdot \vec{p})^2 &= \left(\alpha_x \, p_x + \alpha_y \, p_y + \alpha_z \, p_z \right) \left(\alpha_x \, p_x + \alpha_y \, p_y + \alpha_z \, p_z \right) \\
&= \alpha_x^2 \, p_x^2 + \alpha_y^2 \, p_y^2 + \alpha_z^2 \, p_z^2 + \left(\alpha_x \alpha_y + \alpha_y \alpha_x \right) p_x \, p_y \\
&\quad + \left(\alpha_y \alpha_z + \alpha_z \alpha_y \right) p_y \, p_z + \left(\alpha_z \alpha_x + \alpha_x \alpha_z \right) p_z \, p_x. \tag{20.24}
\end{aligned}$$

Using the results of Eqs. (20.23) and (20.24) in Eq. (20.22) yields

$$\begin{aligned}
&\left\{ E^2 - c^2 \left[\alpha_x^2 \, p_x^2 + \alpha_y^2 \, p_y^2 + \alpha_z^2 \, p_z^2 + \left(\alpha_x \alpha_y + \alpha_y \alpha_x \right) p_x \, p_y \right. \right. \\
&+ \left(\alpha_y \alpha_z + \alpha_z \alpha_y \right) p_y \, p_z + \left(\alpha_z \alpha_x + \alpha_x \alpha_z \right) p_z \, p_x \Big] \\
&- mc^3 \left[\left(\alpha_x \beta + \beta \alpha_x \right) p_x + \left(\alpha_y \beta + \beta \alpha_y \right) p_y + \left(\alpha_z \beta + \beta \alpha_z \right) p_z \right] \\
&- \beta^2 m^2 c^4 \Big\} \Psi = 0. \tag{20.25}
\end{aligned}$$

We require this equation to be equal to

$$\left(E^2 - c^2 p^2 - m^2 c^4 \right) \Psi = 0. \tag{20.26}$$

Comparing terms in Eqs. (20.25) and (20.26), we find the following. Since

$$p^2 = p_x^2 + p_y^2 + p_z^2, \tag{20.27}$$

we see that the second term in Eq. (20.25), that multiplied by c^2, will be equal to p^2 if

$$\alpha_x^2 = \alpha_y^2 = \alpha_z^2 = 1, \tag{20.28}$$

and

$$\begin{aligned}
\left(\alpha_x \alpha_y + \alpha_y \alpha_x \right) &= [\alpha_x, \alpha_y]_+ = 0, \\
\left(\alpha_y \alpha_z + \alpha_z \alpha_y \right) &= [\alpha_y, \alpha_z]_+ = 0, \\
\left(\alpha_z \alpha_x + \alpha_x \alpha_z \right) &= [\alpha_z, \alpha_x]_+ = 0. \tag{20.29}
\end{aligned}$$

The third term in Eq. (20.25), that multiplied by mc^3, is absent in Eq. (20.26). Therefore,

$$\alpha_x \beta + \beta \alpha_x = 0,$$
$$\alpha_y \beta + \beta \alpha_y = 0,$$
$$\alpha_z \beta + \beta \alpha_z = 0. \tag{20.30}$$

Finally, comparing the fourth term in Eq. (20.25) with Eq. (20.26), we see that

$$\beta^2 = 1. \tag{20.31}$$

Thus, the Dirac equation satisfies the relativistic energy relation under the condition that the four relations (20.28)–(20.31) are simultaneously satisfied. Under these conditions, the Dirac equation can be treated as the relativistic form of the Schrödinger equation.

Chapter 21

Systems of Identical Particles

Problem 21.1

Consider a system of three identical and independent particles.

(a) What would be the level of degeneracy if particle 1 of energy $n_1 = 2$ would be distinguished from the other two particles?

(b) What would be the level of degeneracy if the distinguished particle has energy $n_1 = 1$?

Solution (a)

If a single excitation is present in the system of three identical particles, there are three combinations possible of which of the particles is excited, $(n_1 = 2, n_2 = 1, n_3 = 1)$, $(n_1 = 1, n_2 = 2, n_3 = 1)$, and $(n_1 = 1, n_2 = 1, n_3 = 2)$. Thus, for three identical particles, the degeneracy of the single excitation level is three. If the excited particle is distinguished from the other two particles, then there is only one combination possible $(n_1 = 2, n_2 = 1, n_3 = 1)$. Hence, in this case, the level of degeneracy is one.

Problems and Solutions in Quantum Physics
Zbigniew Ficek
Copyright © 2016 Pan Stanford Publishing Pte. Ltd.
ISBN 978-981-4669-36-8 (Hardcover), 978-981-4669-37-5 (eBook)
www.panstanford.com

Solution (b)

If the distinguished particle is in its ground state ($n_1 = 1$), the level of degeneracy would be two, as there are two possible combinations of n_2 and n_3 with the single excitation: ($n_1 = 1$, $n_2 = 2$, $n_3 = 1$) and ($n_1 = 1$, $n_2 = 1$, $n_3 = 2$).

Problem 21.2

Two identical particles of mass m are in the one-dimensional infinite potential well of dimension a. The energy of each particle is given by

$$E_i = n_i^2 \frac{\pi^2 \hbar^2}{2ma^2} = n_i^2 E_0. \qquad (21.1)$$

(a) What are the values of the four lowest energies of the system?
(b) What is the degeneracy of each level.

Solution (a)

The total energy of the two particles is

$$E = E_1 + E_2 = (n_1^2 + n_2^2) \frac{\pi^2 \hbar^2}{2ma^2} = (n_1^2 + n_2^2) E_0. \qquad (21.2)$$

Hence

$$E/E_0 = (n_1^2 + n_2^2) \qquad (21.3)$$

determines the energies of the system.

The first lowest energy level is for $n_1 = n_2 = 1$ at which $E/E_0 = 2$. The second lowest energy level is for either ($n_1 = 2$, $n_2 = 1$) or ($n_1 = 1$, $n_2 = 2$) and the energy of this level is $E/E_0 = 5$. The third lowest energy level is for $n_1 = n_2 = 2$ at which $E/E_0 = 8$. The fourth lowest energy level is for either ($n_1 = 3$, $n_2 = 1$) or ($n_1 = 1$, $n_2 = 3$) and the energy of this level is $E/E_0 = 10$.

Solution (b)

For $n_1 = n_2 = 1$, there is only one wave function Ψ_{11}, so the degeneracy of the first lowest level is one. There are two sets of n's numbers $(n_1 = 2, n_2 = 1)$ or $(n_1 = 1, n_2 = 2)$, which determine the second lowest energy level. The wave functions corresponding to those combinations are Ψ_{21} and Ψ_{12}. Therefore, the degeneracy of this level is two. There is only one set of numbers $n_1 = n_2 = 2$, which determines the third lowest energy level. Therefore, the degeneracy of the level is one. For the fourth lowest energy level, there are two sets of numbers $(n_1 = 3, n_2 = 1)$ or $(n_1 = 1, n_2 = 3)$. Thus, there are two wave functions Ψ_{31} and Ψ_{13} corresponding to those combinations. Therefore, the degeneracy of the fourth lowest energy level is two.

Problem 21.3

Redistribution of particles over a finite number of states

(a) Assume we have n identical particles that can occupy g identical states. The number of possible distributions, if particles were bosons, is given by the number of possible permutations

$$t = \frac{(n+g-1)!}{n!(g-1)!}. \tag{21.4}$$

For example, $n = 2$ and $g = 3$ give $t = 6$. However, this is true only for identical bosons. What would be the number of possible redistributions if the particles were fermions or were distinguishable?

(b) Find the number of allowed redistributions if the particles were:

    (i) Identical bosons.
    (ii) Identical fermions.
    (iii) Non-identical fermions.
    (iv) Non-identical bosons.

Ilustrate this with the example of $n = 2$ independent particles that can be redistributed over five different states.

Solution (a)

Since two fermions cannot occupy the same state, only three redistributions are possible: (1, 1, 0), (1, 0, 1), (0, 1, 1).

If the two particles are distinguishable, then each has three available states and then the total number of redistributions is $3 \times 3 = 9$.

Solution (b)

(i) According to Eq. (21.4), for $n = 2$ identical bosons, there are $t = 15$ allowed distributions over $g = 5$ states. The allowed distributions are

$$
\begin{array}{ccc}
11000 & 01010 & 20000 \\
10100 & 01001 & 02000 \\
10010 & 00110 & 00200 \\
10001 & 00101 & 00020 \\
01100 & 00011 & 00002
\end{array} \qquad (21.5)
$$

where, e.g., 11000 represents the system state in which each of the first and second states contains one particle, while the remaining states contain none.

There are 15 possible distributions of which 10 have the two particles in different states and 5 have the two particles in the same state.

(ii) Two identical fermions cannot occupy the same state. Therefore, the five-system states in the right column of Eq. (21.5) are not allowed. Thus, there are 10 allowed system states for identical fermions.

(iii) For two non-identical fermions, each of the states with two particles in different states, left and middle columns, is doubly degenerated. Therefore, there are 20 allowed system states for non-identical fermions.

(iv) For two non-identical bosons, each of the states with two particles in different states is doubly degenerated. The right column is also allowed for non-identical bosons, so there are 25 allowed system states for non-identical bosons.